THE INTERSTATE HIGHWAY SYSTEM

Henry Moon
Department of Geography and Planning
The University of Toledo
Toledo, Ohio 43606-3390

Copyright 1994

by the

Association of American Geographers
1710 16th Street
Washington D.C. 20009-3198

Library of Congress Card Number 95-3229
ISBN 0-89291-215-4

Library of Congress Cataloguing in Publication Data

Moon, Henry, 1956-
   The interstate highway system / Henry Moon.
     p. cm. -- (Resource publications in geography)
   Includes bibliographical references.
   ISBN 0-89291-215-4 : $10.00
   1. Interstate Highway System--History. I. Title. II. Series.
HE355.M63 1995
   388.1'22'0973--dc20   95-3229
                                                                                   CIP

Publication Supported by the AAG

Original Cartography by Jodie Moon

Contents

FOREWORD v
PREFACE vii

1 ESTABLISHING THE IHS 1

The Legislative Basis 2
Onto the National Agenda 5
The Enabling Legislation 8
Constructing the System 11
Design Criteria 16
Conclusion 19

2 NATIONAL IMPACTS OF THE IHS 23

Network Capacity and Volume 23
Highway Safety 28
The Reduction of Functional Distance 31
Social Impacts of the System 34
Economic Impacts 39
Environmental Impacts 42
Conclusion 44

3 THE IHS AND METROPOLITAN AREAS 45

Locational Conflict in Urban Areas 46
Central Business District (CBD) Impacts 52
Social Impacts 55
Beltway Development 58
One City's Experience 63
Conclusion 67

4 THE IHS AND REGIONAL DEVELOPMENT 69

    The Network and Regional Redistribution 69
    Transport Costs Versus Benefits 72
    The Network and Regional Migration 74
    The Network's Role in Economic Restructuring 77
    Commuting Processes and Patterns 81
    Conclusion 85

5 IHS INTERCHANGES 89

    Interchange Communities 91
    Interchange Morphology 93
    Toward a Model of Interchange Development...................95
    The Interchange Development Model 99
    Interchanges as Regional Growth Poles 100
    An Interchange Community
    in the West: Little America 102
    Conclusion 105

6 EPILOGUE 107

BIBLIOGRAPHY 111

Foreword

Most of the links in the interstate highway system of the United States were built in the 1960s and 1970s. The majority of students in our introductory and advanced geography courses today cannot remember life before the interstate. The industrial park, the regional shopping mall, and the interchange economic zone with its motel, fast food, and gas station franchises are familiar landscapes to children who have grown up outside of central cities, places where interstate construction sometimes destroyed neighborhoods and livelihoods. This monograph provides us with a rest stop where we can pause to review the trip so far.

Given the continuing significance of this system to the geography of the United States today, it is surprising that so little direct attention has been paid to it by geographers and other social scientists. This monograph offers an overview of the development of the IHS and research from the last several decades on the impact of the system regionally and locally. As our economy evolves, there are many questions that can be raised about the IHS. What are the implications, if any, for the development of free trade zones in the Western Hemisphere for the system? What will be happening to the first-generation suburban shopping malls as they age and face competition from new centers? Will IHS interchange communities become new types of urban settlement? We hope the information presented here encourages a new generation of scholars to engage in research on our transportation systems, both to explain why they are the way they are and to shape public policy on what they should be.

Ellen K. Cromley
Robert Cromley
Co-editors
Resource Publications Series

PREFACE

The subject of this book is the Dwight D. Eisenhower System of Interstate and Defense Highways, also known as the interstate highway system (IHS). Every U.S. citizen and visitor has some experience with this highway network recognized as the largest public works project ever undertaken. One could argue, however, that despite our first-hand knowledge few recognize the scope of the economic, social, environmental, and geographic impacts wrought by the development of the system and its antecedents during this century. Many would not even be aware of the system's official name.

The objective of this monograph is to provide readers with a broad introduction to the IHS. The forces in the private and public sectors contributing to its development as well as the system's role in reshaping form and function in the geography of the U.S. are covered in this overview. I rely heavily on maps, graphs, and tables to document the scope of the network and its role in spatial, social, and economic change. In addition, readers are urged to investigate on their own the depth and breadth of the many scholarly efforts to describe specific aspects of the system, for the literature is rich with works on individual chapters of the IHS story.

Throughout this monograph, the terms "system" and "network" are used often in referring to the IHS as a complex set of connectors spurring sometimes contradictory patterns and processes. As I wrote in the International Geographical Union's *Geographical Snapshots of North America* in 1992, the IHS "segregates and integrates, serves as a facilitator and a barrier, decentralizes and coalesces, transcends and limits" (Moon 1992, 427). Consequently, the presentation here describes more than the simple links and nodes of the geometric network but also considers the complex organism that grows beyond the bounds of its creators. The IHS is more than a project, more than a set of roads. It is a combination of political will, industrial drive, engineering ferocity, governmental bullying at its worst, and, penultimately, an expression of American freedom.

I would like to acknowledge the support of my colleagues in the Department of Geography and Planning at The University of Toledo for their support throughout this project. Their generous moral and editorial support is appreciated more than they know. In particular, I would like to express my gratitude to Jim Nemeth, an unselfish and willing reader, and Bill Muraco. In addition, I would like to recognize The University of Toledo Office of Research and my editors, Ellen and Bob Cromley. Finally, I would like to thank the members of my family for their infinite patience and support because without them this project would have been impossible. And, most of all, I would like to thank my life partner Jodie for being my sole inspiration and this project's chief cartographer.

Henry Moon
Toledo, Ohio
September, 1994

1

Establishing the IHS

> Let us then bind the Republic together with a perfect system of roads and canals. Let us conquer space...
> (John C. Calhoun quoted in Patton 1986, 25)
>
> The highways of America are built chiefly of politics whereas the proper material is crushed rock or concrete. (Carl Fisher quoted in Patton 1986, 53)
>
> Like television, American highways are a national network, a mass medium. (Patton 1986, 20)

When historians and historical geographers examine the changes that occurred in the United States during this century, they might well list the development of the Interstate Highway System (IHS) among those having the most far-reaching impact on our nation. In addition to the impact of its actual construction, development of the system contributed to a shift of political power from the state to the federal level and transformed the spatial organization of our economic and social institutions. Garrison, one of the first geographers to study the IHS (1959), emphasized the system's ability to change some of our society's preeminent geographical patterns. He cited possible impacts on central places, retail location, residential development, shopping behavior, and the provision of basic services. Fifteen years later, Garrison (1974) reported that the IHS was "inducing" changes in the relative locations and economic prosperity of urban places and documented the shifts in urban service areas and service function locations within cities. He concluded that many of the traditional geographic perspectives and concepts developed to explain transportation networks failed to adequately address the IHS and its impacts.

The development of transportation systems like the IHS has contributed to areal specialization at larger scales. Ullman (1956) contended that increased regional interaction (circulation) drives the formulation of networks. When regional complementarities exist, trading regions interact using transportation systems as the media. But, the spatial organization of activities changes through time with new intervening opportunities arising, often at key network intersections or interchanges. Major nodes emerge at these locations as agglomerative forces are attracted to transportation intersections. Network evolution frequently operates to enhance transferability by altering the friction of distance. The history of the IHS exemplifies these principles and a review of its impacts illustrates how the space economy of the U.S. and its people have changed as a result.

The Legislative Basis

When Congress passed *The Act To Regulate Commerce* in 1887, few probably realized its long-term significance for the geography of the United States. This legislation was written with the specific objective to regulate the country's burgeoning railroad industry and created the the nation's first independent regulatory agency, the Interstate Commerce Commission (U.S. Congress 1887). The Act had important implications for future developments in transportation. It marked the initial effort to establish a national transportation policy in the U.S. and outlined the general relationship between the federal government and transportation. Because the Act failed to establish a national policy, an era of *"trial and error regulatory planning in which Congress used a step-by-step approach to patching here and reinforcing there"* began (Mertins 1972, 10).

In 1893, the U.S. Office of Road Inquiry was created within the Department of Agriculture (Patton 1986, 5). In spite of the fact that few automobiles existed at the time, there were widespread public calls for governmental intervention to improve the poor road system, then in its infancy. In 1895, only four automobiles were registered in the entire U.S. but the number grew to approximately 8,000 by 1900 and to 458,000 by 1910 (Mertins 1972, 12). It is important to note that most of the calls for improved roads came from rural regions and the Congressional response established an early locational bias in the national approach to highway planning, funding, construction, and regulation. The fact that the new department opened under the purview of the Department of

Agriculture reflects this early governmental attitude toward road development in rural areas. The Office of Road Inquiry conducted the first road census in 1904 reporting that only 7 percent of U.S. roads were surfaced and that only one mile of improved road existed per 492 citizens (Flink 1990, 5).

Although the Office of Road Inquiry was the designated governmental road transportation agency, the driving force behind the most significant single piece of federal highway legislation came from organizations outside government. *The Federal-Aid Road Act of 1916* was forced onto the Congressional agenda by a group of private associations collectively known as the *"Good Roads Movement"* that were dedicated to widespread road access (U.S. Congress 1916; Flink 1990, 170). Unlike the effort to construct the National Road 100 years earlier (Vance 1986, 495), this movement originated in the private sector.

Among the nearly two dozen privately funded organizations calling for more and better roads, the most effective was the Lincoln Highway Association (Vance 1986). It was formed in 1913 by automobile, gasoline, and construction industry executives and was originally named the Coast-to-Coast Rock Highway Association. The name change was calculated to draw on the popular sentiment surrounding Lincoln's name. The Association was seeking to attract broad political support for roads, in general, and the Association's proposed transnational Lincoln Highway to run from New York City to San Francisco, in particular. The planned route of the Lincoln Highway was similar to the routes of early pioneer trails, the National Road, and today's I-80. The Lincoln Highway Association never saw the completion of its namesake route even though the roadway garnered a great deal of national attention when the first segment was opened in Illinois in 1915. Although the would-be highway never grew beyond a set of isolated, predominantly rural segments, the project was successful in focusing public attention on the need for a national network of quality roads.

Another private highway advocacy group, the National League for Good Roads, was also instrumental in the movement to build more and better roads. The League consisted of farmers, businessmen, transportation advocates, and early transportation analysts (Mertins 1972). This group held several national Good Roads Conventions beginning in 1893 and called for county, state, and federal government support of road improvement and construction.

These efforts and those of a number of other organizations were important in initiating governmental intervention in surface transportation (Vance 1986). Just ten years after the roads census, the evolving U.S. road system featured 257,293 miles of surfaced roads including 75,400 miles paved with macadam, 1,591 miles paved with brick, and 2,349 miles paved with concrete (Flink 1990, 170). Ironically, the Good Roads Movement was heavily backed by both bicycle advocates and the railroad industry. It is important to note the dominant role still played by railroads during this period of U.S. history. As late as 1920, 84 percent of intercity freight and 85 percent of all passenger travel on public carriers moved by rail (Mertins 1972, 13).

The Federal-Aid Road Act of 1916 marks the beginning of Congressional interest in developing a nationwide highway network and provides the legislative and constitutional basis for the IHS (U.S. Congress 1916). Through the Act, Congress authorized and appropriated $75 million to be matched equally by individual states for the construction of rural "post" roads (Dearing and Owen 1949, 106). This fiscal arrangement established a funding precedent that later spread to other modes of transportation and eventually beyond transportation to other programs. The Act required the formation of highway departments in participating states and federal instruction to *"develop management and construction standards acceptable to the federal government"* (Mertins 1972, 12).

A 1921 amendment to the Act required individual states to designate a system of interstate and intercounty routes deemed eligible for federal road aid. Mileage could not exceed 7 percent of a given state's total road mileage (Dearing and Owen 1949, 106). Given a lack of revenue, especially in the west, states were strapped to find the necessary matching funds. Lack of matching funds prompted the Oregon legislature to enact the first gasoline tax in the U.S. in 1919 (Flink 1990). During the next ten years the taxation of gasoline spread to every state and the District of Columbia with the citizenry apparently welcoming this new form of taxation. According to Flink (1990, 171), one Tennessee official asked *"who ever heard, before, of a popular tax... never before in the history of taxation has a major tax been so generally accepted in so short a period."* Apparently, the citizenry had become willing to fund their newfound freedom of the road.

Prior to 1928, the focus of Congressional road policy was the rural United States. However, the passage of a key amendment that year, also enabled federal urban highway aid (U.S. Congress

1928). This first step was "half-hearted" but it represented a radical departure in terms of the geographical orientation of Congressional funding (Smerk 1965, 123). Federal road improvement funds specifically designated for only the widest city streets became available to communities of 2,500 or more. During the Great Depression, President Franklin D. Roosevelt used the construction and improvement of urban streets to increase employment.

An important related development was the use of federal funds for the construction and improvement of regional "farm-to-market" roads (Smerk 1965, 124). *The Emergency Relief and Construction Act of 1932* appropriated and allocated funds specifically for the enhancement of secondary and rural-to-urban feeder roads (U.S. Congress 1932). One year later *The National Industrial Recovery Act* forced federal road intervention deeper into urban America by establishing the funding of highways "into and through" cities (U.S. Congress 1933; Smerk 1965, 124). The urban geography of the United States was forever changed by the connecting of cities and their peripheries via these improved roads. Functional metropolitan regions, each featuring a distinct urban core and a surrounding hinterland, were reinforced by legitimate road systems for the first time. This activity took place under federal auspices largely without local funding or input.

Onto the National Agenda

After the Great Depression, government and industry seemed increasingly interested in a national highway system. Leaders across the country were aware of the progress being made on Germany's *Autobahn* and were convinced of the role that superhighways would play in the looming land-based war (Banks 1984, 76). Adolph Hitler postulated in his first major speech that *"a nation is no longer judged by the length of its railways but by the length of its highways"* (Flink 1990, 262). Hitler's *Autobahn* became *"the first integrated network of express highways in the world...a model for all future highway construction"* (Flink 1990, 262). By 1942, the *Autobahn* was comprised of 1,310 miles of integrated freeways and spurred international competition in road planning and construction (Vance 1986, 513).

The specific idea behind and label for a *"National System of Interstate and Defense Highways"* for the U.S. can be traced to a 1935 meeting between President Roosevelt and representatives from the Bureau of Public Roads (Banks 1984, 76). The first official step toward development of a federally mandated network was made when Congress authorized the Bureau (soon to become the Public Roads Administration) to undertake a study of an integrated system of toll roads to stretch across the United States. The study, entitled *Toll Roads and Free Roads*, was released in the Spring of 1939 and, raised three primary points (Banks 1984). First, insufficient transcontinental traffic existed to finance highway construction via tolls. Second, a freeway system capable of meeting future peacetime traffic needs as well as those of the national defense was needed. Finally, a system of 26,700 freeway miles (not the 14,336 miles of toll roads proposed by Congress) planned through a cooperative effort between the Public Roads Administration (PRA), the War Department, and individual states would best serve the long-term national interest. The highway network recommended in the report would be heavily oriented toward metropolitan areas and that bias helped garner popular support. For the first time, an "official" report called for establishment of a national highway network.

During development of this report, a number of ideas with geographical implications were proposed. For example, eminent domain power -- the taking of private property for the common good -- under the control of a single federal agency was suggested by President Roosevelt as a means to generate highway funds (Patton 1986). After seeing the profits generated by railroad companies as they spawned economic development through increased accessibility, Roosevelt supported a plan in which the federal government would take key pieces of real estate along new highways (Vance 1986). Later the government would sell these properties on the free market to the highest bidder, generating large profits to reimburse highway construction. The "socialist" overtones of the President's proposal resulted in its rejection by Congress and among the public at large (Patton 1986, 74). Rejection meant that the "limited access" concept would win over Roosevelt's idea of unlimited access and linear, development prone corridors along freeways (Vance 1986).

The Transportation Act of 1940 established for the first time specific goals for a national transportation policy (U.S. Congress 1940). The goals involved the *"development, coordination, and preservation of a national system to meet the needs of commerce, the Postal Service, and national defense"* (Mertins 1972, 35). In addition to loosening federal regulations on railroads and increasing those on water-based carriers, this legislation mandated federally funded transportation research. This Act and the new focus on research increased Congressional intervention into highway transportation. For national legislators of the time, roads became the object of legislation that railroads had been for most of the previous century. However, public policy discussion of a national highway network came to a virtual halt with the outbreak of World War II.

Following the end of the War, Congress turned its attention back to highways. Again, an organization of firms that stood to profit from construction of *Autobahn*-like superhighways operated in the background, nudging lawmakers toward greater federal involvement (Flink 1990). This group, known as the "Road Gang" was more secretive than its earlier counterparts but the thrust was much the same. Meeting since 1942, the group was determined to steer post-war funding toward an interstate highway network. The group *"had some 240 members, including representatives of the automobile manufacturers and dealers, automobile clubs, oil companies, truckers and the Teamsters Union, highway engineers, and state highway administrators"* (Flink 1990, 371). Their efforts in conjunction with memories of the *Autobahn* resulted in a continuation of the pre-war Congressional push to develop national superhighways. Consequently, the Public Roads Administration published another major report that became the blueprint for the new network. *Interregional Highways*, published in 1944, called for the creation of an integrated system having approximately 39,000 miles of limited access freeways (U.S. Department of Transportation, Federal Highway Administration 1976c, 468). Routes would connect major military installations and cities of 300,000 or more (Banks 1984, 77). Other factors to be considered in the planning of the network were population distribution, manufacturing activity, employment patterns, existing traffic flow patterns, and topography (Banks 1984). It was generally thought that this system could be developed by upgrading existing roads and streets (Vance 1986).

Based on this detailed report, Congress moved to enact *The Federal Aid Highway Act of 1944* authorizing creation of the IHS. The system's stated purpose was to connect by routes, as direct as practical, the principal metropolitan areas, cities, and industrial centers to serve the national defense, and to connect at suitable border points with routes of continental importance (U.S. Department of Transportation, Federal Highway Administration 1976c, 468). However, no funds were allocated for construction of the proposed network.

Discussion continued and the PRA's *Highway Needs of the National Defense* report, published in 1949, argued for the accelerated completion of a superhighway system to meet the growing needs of the military and, as a secondary benefit, to enhance the civilian economy (Banks 1984). Unlike the 1944 report, this study called for a system of predominantly new routes strategically located between large urban areas and major military bases. The cost estimate for the system was an astounding $11.3 billion (U.S. Department of Transportation, Federal Highway Administration 1976c, 469). More significant was the report's recommendation to solve the funding problem that had prevented construction to date by shifting a greater share of the fiscal responsibility to the federal level (Banks 1984). From the issuance of this report to the mid-1950s, Congress reviewed a number of additional high profile and widely respected reports recommending alternative paths for developing the system.

Once again, the private sector stepped in to provide the impetus for the legislation that would make the system a reality. Ford Motor Company, enjoying production and sales increases of proportions not experienced up to the time, commissioned a book calling for citizen involvement in the movement for better roads (Patton 1986). *Freedom of the American Road* (Lass 1956) detailed citizen participation in planning new roads, highlighting successful case studies of such activities and listing the benefits of highways. The book equated the push for road building with good citizenship. Henry Ford II wrote that *"we Americans always have liked plenty of elbow room -- freedom to come and go"* and a citizen movement toward better roads would represent *"democracy in action and resourcefulness"* (Patton 1986, 89). Given public interest in highway construction and a known tolerance for gasoline taxes on the part of consumers, Congress finally moved toward funding a national highway network. No single interest can claim credit for the action. It took the combined efforts of the military establishment, industrialists, and members of the transportation community, primarily engineers.

The Enabling Legislation

The Federal-Aid Highway Act of 1956 and the *Highway Revenue Act of 1956* were signed into law by President Dwight D. Eisenhower on June 29, 1956 (U.S. Congress 1956a; U.S. Congress 1956b). Unlike the many highway acts and amendments to date, these dual acts not only authorized a program but appropriated the necessary funds for project implementation. The passage of *The Federal-Aid Highway Act of 1956* followed a period in U.S. history when the number of privately owned automobiles dramatically increased and a number of states were building toll roads to carry increasing traffic. The presence of a popular military figure in the White House and the prevailing national mindset following World War II culminated in the action to build the IHS (Patton 1986). President Eisenhower had firsthand experience with the German *Autobahn* and he drew on that experience to implement the network. Explaining Eisenhower's perspective:

> *The Interstate program was the last New Deal Program and the first space program, combining the economic and social ambitions of the former with the technological and organizational virtuosity, the sense of national prestige and achievement, of the latter. The planners of the system expanded the public works aspect of highway building into a vision of Keynesian pump-priming and economic fine-tuning, directed under Eisenhower through private industry. They linked the economic vision to a dubious one of national defense. They adapted the aesthetic - and implicitly the social - values of the American parkway tradition to the new suburbia. They looked back to the German Autobahns for standards of beautiful and efficient engineering. And they answered a public demand for the realization of the utopian, technological future outlined by the streamlining visionaries of the World's Fairs and Sunday supplements. (Patton 1986, 85-86)*

Many observers have concluded that the President was proud of his role in initiating the system but disappointed by its outcome. During a drive from The White House to Camp David in 1959, Eisenhower was reportedly surprised by interstate construction in an urban area (Patton 1986, 98). His impression of a predominantly rural interstate system "astonished" those aware of it because the President had supposedly worked with detailed system maps indicating routes through cities. In addition, he had relied on the political support of many urban congressmen and mayors to push the system through Congress. An even greater contradiction to his misconception was his own argument for a highway system capable of evacuating America's urban population given a nuclear attack. In any case, he minimized this landmark achievement of his administration in his memoirs (Patton 1986).

Title I of the dual legislation *The Federal Highway-Aid Act of 1956* contains a set of key provisions outlining the interstate system. These provisions established an official title, The National System of Interstate and Defense Highways, and lengthened the system's mileage to 41,000. The provisions also established construction standards as well as location and design criteria and specified projected 1975 traffic levels as the planning threshold. The traditional biennial funding authorization was discarded by extending federal funding into the late 1960s and setting up a complex funding formula intended to guarantee financial support throughout construction through the Highway Trust Fund. For the first time, the proportion of federal funding responsibility was increased to at least 90 percent (U.S. Department of Transportation, Federal Highway Administration 1976c, 472).

Title II, *The Highway Revenue Act of 1956*, established a long-lasting precedent in terms of the fiscal relationship between Congress and the individual states (U.S. Department of Transportation, Federal Highway Administration 1976c). It provided the missing link between good intentions and actions that had prevented the formation of a national highway network. Title II enabled funding of the prescribed IHS through federal excise taxes on system users. Collected user taxes would flow through the Highway Trust Fund to participating states under the 90/10 percent funding split. Relying primarily on gasoline taxes (the most prevalent user-based excise tax), Congress had finally moved away from the traditional 50/50 fund matching relationship it had long held with individual states. For the first time, federal highway funds would come from a revenue source other than the general budget; the *"pay as you go"* concept was born.

By 1990, the trust fund received a 9.1 cents per gallon gasoline tax, 15.1 cents per gallon diesel fuel tax, graduated tax on tires over 40 pounds, 12 percent excise tax on heavier trucks and trailers, and a user tax on trucks weighing over 55,000 pounds (The Road Information Program 1990, 26-27). Since the first day of its collection, the motor fuel portion of the tax package has been the dominant contributor (Figure 1.1). The proportion of the Trust Fund coming from these fuel taxes, however, has been variable. The motor fuel tax share of the trust fund decreased from 97 percent in 1957 to just 57 percent in 1980 returning to 71.6 percent by 1987 (The Road Information Program 1990, 27). Since Fiscal

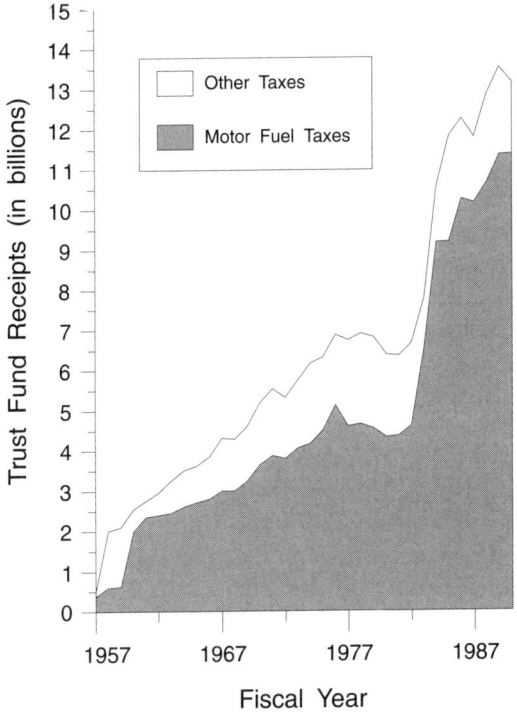

Figure 1.1. Federal Highway Trust Fund receipts from 1957 to 1990. Source: U.S. Department of Transportation, Federal Highway Administration 1984c and 1991b.

Year (FY) 1957, the Highway Trust Fund has grown to represent the expanding power that the federal government holds over individual states.

Constructing the System

For a century, groups with varied transportation goals and interests called for a greater federal role in developing all modes of transportation. Literally thousands of agencies, firms, and individuals have been directly involved in its development and its diffusion across the U.S. With every election, piece of highway legislation, economic cycle, and war, the system changed. It was altered not only in purpose or intent but in actual design. Early goals to improve existing roads were replaced by ones to fund all new construction. What was initially to be a system of toll roads became one of freeways funded through taxes. The whole concept of limited access, four-lane superhighways occurred during, not prior to, the planning process. The IHS was never truly planned but allowed to evolve over a relatively long period of time. Consequently, its history, design, purpose, and eventual outcome reflect inconsistency and conflict.

The original layout of the IHS is probably most similar to the road network represented in the Pershing Map produced in 1922 (Figure 1.2). In response to a request from the Bureau of Public Roads, General John J. Pershing, then General of the Armies, constructed a map entitled *"Project for the Development of National Highways of the United States"* (U.S. Department of Transportation, Federal Highway Administration 1976c, 142). This map included all those routes deemed important by the War Department (which later became the Department of Defense). War Department cartographers and planners drafted a set of routes to be used for large scale evacuations of U.S. cities and another set of more isolated routes having primarily military significance (Vance 1986). The Pershing Map featured over 78,000 miles of strategically located routes, well within the mileage cap of 200,000 miles set forth in a 1921 amendment to *The Highway Act of 1916*.

Congress and federal highway planners eventually pared Pershing's mileage by almost half but his map remains the primary

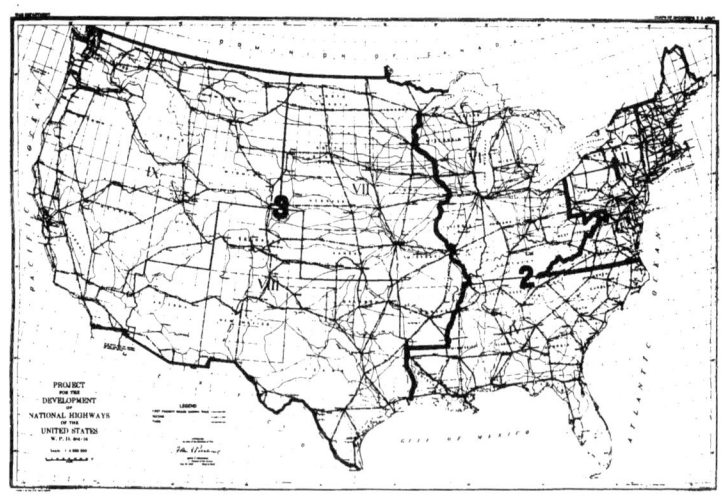

Figure 1.2. The Pershing Map. Source: U.S. Congress 1949.

planning document behind the IHS. A comparison of the Pershing Map with the evolving IHS reinforces its role as the prototype for the network (Figure 1.3). From the outset, highway projects under construction by 1960 and existing toll roads that were folded into the system incorporated elements of the Pershing Map (Figures 1.2 and 1.3a). The Commissioner of Public Roads reaffirmed the role of the Pershing Map in shaping the IHS by stating that:

> *It seems apparent that the system of highways of greatest importance to the national defense, having its genesis in the routes of the 1922 Pershing Map, ... has now culminated in the national system of interstate highways. (Dearing and Owen 1949, 119)*

Perhaps more important, a mutually beneficial relationship between the two national governmental agencies in charge of the military and roads was established immediately following World War II. National defense became one of the governing criteria, if not the most important consideration, in determining the nature and content of federal transportation policies (Mertins 1972).

When it adopted *The Federal-Aid Highway Act of 1956*, Congress refrained from a strong role in actually planning the location and design of the IHS. Congress "permissively modified" the language of the 1944 legislation that called for a network to connect the country's predominant metropolitan areas (U.S. Department of Transportation, Federal Highway Administration 1976c, 473). In the 1956 act, Congress, given its objective of prompt system completion, prescribed the following planning perspective:

> *Insofar as possible in consonance with this objective, existing highways located on an interstate route shall be used to the extent that such use is practicable, suitable, and feasible, it being the intent that local needs, to the extent practicable, suitable, and feasible, shall be given equal consideration with the needs of interstate commerce" (U.S. Department of Transportation, Federal Highway Administration 1976c, 472).*

Several key planning elements were added to the 1956 legislation. Because the country had minimal experience with superhighways and their problems (only 2,000 miles of new toll roads were in place at the time), legislators were cautious, according to the U.S.

14 Establishing the IHS

a. Source: U.S. Department of Commerce, Bureau of Public Roads 1960a.

b. Source: U.S. Department of Transportation, Federal Highway Administration 1970a, December 31, 1970.

Figure 1.3. The evolving IHS.

Department of Transportation, Federal Highway Administration (1976c, 472). Commercial activities were barred from location on interstate routes and/or from access to interstate highway rights-of-way. The air space above and the subterranean space below interstate routes could be used for parking. Vehicle weight and load limitations were indirectly imposed on states when they

c. Source: U.S. Department of Transportation, Federal Highway Administration 1980a, June 17, 1981.

d. Source: U.S. Department of Transportation, Federal Highway Administration 1990a, September 1, 1992.

sought Highway Trust Fund dollars. Public hearings in and around metropolitan areas were mandated to facilitate the exchange of ideas concerning route location (hearings were held after the local routes were planned so in effect they served as public presentations, not as open forums). Federal eminent domain proceedings were allowed to circumvent local and state laws capable of delaying land acquisition. In addition, Congress created a mechanism for including certain toll roads in the IHS if they met minimum design criteria. In 1957, 2,102 miles of toll roads were accepted into the system under the agreement that they would eventually become interconnected (Mitchell 1958, 324). By 1992, the IHS would contain 2,237 miles of toll roads -- roughly 5 percent of its total mileage (U.S. Department of Transportation, Federal Highway Administration 1992a, 1).

The actual planning of the system was left to the Bureau of Public Roads and the 48 state highway departments through the American Association of State Highway Officials (AASHO). Later, the network that they designed would become the basis for a planned intercontinental network of superhighways. The North American and South American Motor Road Networks jointly form the Pan-American Highway System (Figure 1.4) (Larson 1964).

Design Criteria

As charged by Congress, AASHO developed the original set of design standards for the IHS. Their general objectives were to design a network that would be a *"credit to the Nation"* and promote its *"economic welfare and defense"* (American Association of State Highway Officials 1956, 1). The criteria established for the original construction and planning of the system stipulated that all highway links would include a minimum of four traffic lanes and that the links would function as two separate one-way roads. In addition, the entire system would be limited access to *"insure [highway] safety, permanence, and utility and with flexibility to provide for possible future expansion"* (American Association of State Highway Officials 1956, 1). Also, the highways would be designed to promote relaxed travel along pleasing roadways. A set of specific design standards evolved from these basic objectives and addressed all aspects of the planned network including anticipated traffic levels. In effect, the drafters of these design standards held constant the highly variable physical and social characteristics of the U.S.

The design year for a segment was established as the year 20 years in the future from the anticipated year of a segment's actual construction. "Design year" is a concept used in transportation planning to anticipate and build for future traffic volumes. In this case, volumes were to be projected two decades ahead. Projected traffic volumes to be used in planning were those of the 30th

Figure 1.4. The Pan-American Highway System. Source: Larson 1964. Reprinted with permission of Eno Transportation Foundation, Inc.

highest hourly volume of the design year. This number represented a compromise between those wanting to plan for peak traffic and those arguing for a more conservative and less-costly lower planning volume. The 30th highest hourly volume became known as the *"design hourly volume"* or DHV.

Controlled access to the network was accomplished through right-of-way acquisition. Although urban right-of-way margins were not specified by planners because of their potentially high cost, rural margins were set at 150 to 300 feet depending on lane number and the presence of a frontage road. Access control was extended beyond the highways to include ramps, terminals, driveways, and some frontage roads. The area over which Federal control extended to provide for access to the network was linear, extending down intersecting road corridors from the intersection with the highway. The extent of this control was set at 100 feet in urban areas and 300 feet in rural areas (American Association of State Highway Officials 1956, 2). All at-grade highway intersections and railroad crossings were to be eliminated.

The design speeds established by AASHO planners were functions of local topography with little attention paid in the early designs to variable population and/or traffic densities. The original design speed criteria were 70 mph in relatively flat areas, 60 mph per hour in "rolling" areas, 50 mph in mountainous areas, and at least 50 mph in urban areas (American Association of State Highway Officials 1956, 3). Highway curvature, elevation, sight distance, and grade would be controlled by the specific design speeds. Gradient maximums were set at 3 percent on 70 mph segments, 4 percent on 60 mph segments, and 5 percent (extendable to 7 percent) on 50 mph segments. When AASHO planners established the IHS design speed criteria in 1956, they could not have been aware of the controversy over speed limits and legislation that would emerge in the future.

A minimum lane width of 12 feet was established for the network, expanded to include "climbing" lanes in mountainous areas (American Association of State Highway Officials 1956, 3). In urban or in mountainous areas, medians would ideally be at least 16 feet wide. Realizing the difficulties associated with achieving this ideal, planners insisted on an actual four foot minimum. Median widths in flat and rolling rural areas were set at 36 or more feet. Minimum shoulder width was set at six feet in mountainous areas and 10 feet in all others while minimum bridge clearance was set at 16 feet in rural areas and 14 feet in urban areas. Like the standards addressing speed, the minimum width design standards would pose difficulties in developing the network throughout its construction and its reconstruction.

Route numbering was also carried out through the AASHO. Their design guidelines specified that even-numbered routes run east-west (for example, I-80) and that odd-numbered routes run north-south (for example, I-35) (U.S. Department of Transportation, Federal Highway Administration 1976c). Trunk lines have one- or two-digit designations (for example, I-8) while longer, relatively evenly spaced routes have numerical designations ending in zero or five (for example, I-40). To avoid confusion with existing U.S. routes, the lowest numerical designations were located in the south and west (for example, I-10). Urban bypasses and loop roads have three-digit designations based on the main route number and an even number prefix (for example, I-405). Radial and spur routes also have three-digit designations based on the main route number but with an odd number prefix (for example, I-110).

Today, the system connects the contiguous 48 states with 58 one- or two-digit designated trunk lines (Figure 1.3). Twenty-seven of these are even numbered (running east-west) while 31 are odd numbered (running north-south). Hawaii now has three routes (H1, H2, and H3) while Alaska and Puerto Rico have similarly designated routes that are not part of the IHS. A total of 244 local/regional beltways, spur lines, and connectors exist in the network. The IHS's five longest routes (I-10, I-40, I-70, I-80, and I-90) extend from coast to coast and are each in excess of 2,000 miles in length. I-90 is the longest single route, stretching 3,081 miles from Boston to Seattle. North-south routes dominate the list of those between 1,000 and 2,000 miles long (I-5, I-15, I-20, I-25, I-35, I-75, I-94, and I-95). The shortest route on the IHS is I-878, a 0.7 mile connector in New York City.

Conclusion

From the perspective of the 1990s, the IHS has emerged from public and private efforts extending back to the late 19th century (Table 1.1) Even among those most involved in formulating IHS policy, few could probably fully realize the long-term implications of their actions as the system continued to evolve. States were encouraged to take swift advantage of available federal funds by accelerating IHS planning and construction, a response that shifted the state focus away from local routes (Mertins 1972; Patton 1986). The intercity transportation of people and goods would be dominated by the automobile, bus, and truck while intracity transportation would be dominated by the family car. As city centers declined with respect to their surrounding suburbs, the form and function

TABLE 1.1
Top Ten Years in the IHS Chronology

Year	Major Event in IHS Development
1887	Passage of The Act to Regulate Commerce
1893	U.S. Office of Road Inquiry Established
1916	Passage of The Federal-Aid Road Act of 1916
1922	Pershing Map of National Defense Highways Circulated
1939	Toll Roads and Free Roads Report Issued
1944	Interregional Highways Report Issued Passage of the Federal-Aid Highway Act of 1944
1949	Highway Needs of the National Defense Report Issued
1956	Passage of the Federal-Aid Highway Act of 1956 IHS Construction Began
1976	Funds Authorized for 3R (Resurfacing, Restoration, and Rehabilitation) Work on the IHS
1990	IHS Officially Designated "The Dwight D. Eisenhower System of Interstate and Defense Highways"

of American metropolitan areas changed. The post-World War II trend of suburbanization was accelerated as people and industry migrated away from city centers.

Federal funding practices, planning, and decision making were altered in favor of one mode, the automobile, outside of any overall planning strategy. The focus was clearly placed on the country's interstate network but what about the others? A *"negative federal energy policy"* was established that tied increased gasoline consumption to the number of dollars available for highway construction because the gasoline tax was set per gallon and not as a percentage of price (Patton 1986, 94). Funding was based on a state's IHS length in miles, a function of area, rather than population, directing massive federal outlays to states like Texas and California. The economic boom of the area later recognized as

Conclusion

the Sunbelt was enabled, consistent with Ullman's (1956) views on the role of transportation in regional development.

A national precedent was established placing circulation at a higher priority than integration -- the IHS was built to facilitate movement, not to bind the country together. Patton (1986, 95) writes that *"to engineers, this meant no more than keeping traffic moving; to the interest groups, it meant keeping the flow of their products moving; to the politicians, it had economic, military, patriotic, and almost spiritual implications."* The IHS would become the largest public works project ever undertaken (U.S. Department of Transportation, Federal Highway Administration 1976c). The network would reshape the human and physical geography of the United States.

2

NATIONAL IMPACTS OF THE IHS

> *Highways have been a prime contributor to the dispersion of urban America. The stimulation provided by the federal programs for highways has done as much as any other single force to bring the United States into the automotive age. Change has never been an unqualified bargain; it exacts a stiff price... the price of the automotive age has been congestion. (Smerk 1965, 138-139)*

> *Highways are much more than a means of transportation. They come as close as anything we have to a central national space. (Patton 1986, 21)*

Automobiles and other highway vehicles have altered the physical and human geography of the United States more so than perhaps any other form of transportation. From the advent and widespread diffusion of Henry Ford's family car, the form and function of the country have changed. IHS development epitomizes the public and private sector desire to ensure "American automobility" (Flink 1990, 372). The United States has become automobile and IHS dependent, with most intercity passenger travel occuring by automobile and an increasing percentage of intercity freight tonnage moved by highway (U.S. Department of Transportation, Federal Highway Administration 1984c). This chapter describes the volume of travel on the IHS and examines its national impact.

Network Capacity and Volume

Reasonable national estimates of interstate traffic flow are conspicuously absent in published data about the IHS. Very little comprehensive data on interstate travel have been collected since the 1939 publication of *Toll Roads and Free Roads*. Although most states collect volume data by means of roadside traffic counts, there is no coordinated federal effort to collect, analyze, or map interstate travel. Other than the annual Federal Highway Administration entry on estimated vehicle miles traveled (VMT), there is little additional data. Although the estimated VMT is broken down in terms of urban versus rural travel and by highway type, it is not available by state or by route. Based on what we do know about interstate flows, however, some important conclusions about the system's impacts related to network capacity can be drawn.

Early on, there was a great deal of debate surrounding the capacity of the IHS. Highway capacity is the maximum number of vehicles that a segment can safely and efficiently carry at a given time. In the authorization/appropriation legislation of 1956, Congress had targeted 1975 as the date of completion and early planning was geared toward that year. In 1963, however, Congress mandated designation of a "design year" two decades into the future, accommodating both construction delays and the general lack of concensus on likely traffic levels. Design year designation was important because it extended the focus of early interstate planning efforts beyond 1975. Consequently, the mileage planned and built prior to 1963 was designed to accommodate 1975 traffic while routes planned after 1963 were designed to accommodate traffic estimates for a design year 20 years hence.

An examination of the system's status in 1963 reveals how quickly the system was coming on line. Of the 1992 network mileage (42,795 miles), 38.7 percent was open to traffic at the end of 1963 and another 39.8 percent was classified as in progress (U.S. Department of Transportation, Federal Highway Administration 1992a; U.S. Department of Commerce, Bureau of Public Roads 1964a). Collectively, these data indicate that 78.5 percent (33,594 miles) of the system in place in the early 1990s was designed to carry 1975 traffic levels. Not surprisingly, congestion was an immediate problem in and around some urban areas. According to a 1963 study, 21.7 percent of the network's mileage, including toll roads incorporated into the IHS, were capable of carrying 1963 traffic volumes but would require upgrading to bring them into alignment with standards as traffic volume increased (U.S. Department of Commerce, Bureau of Public Roads 1964a, 1). Because of this drawn out design period, the time needed to complete impact assessment studies, and construction delays, much of the system's urban mileage was brought on line at or near capacity.

Major discrepancies between Federal Highway Administration projections and actual traffic volumes have resulted in episodes of congestion, primarily in urban areas, unmatched in our road transportation history. These same differences also resulted in the dramatic under-utilization of most rural segments. The IHS represents only 1.2 percent of U.S. highway mileage but carries 22.3 percent of all vehicle miles traveled (Table 2.1). The urban share of total interstate mileage is 25.5 percent, but it carries 58.1 percent of the network's traffic, a flow of some 278,404 million VMT in 1990 (U.S. Department of Transportation, Federal Highway Administration 1991b, 122-123, 195). Although it is not uncommon for average daily traffic (ADT) counts to fall below 10,000 on some rural interstate segments, traffic counts taken around some of the nation's major metropolitan areas routinely exceed that level by a factor of 25. Los Angeles, Atlanta, Chicago, and New York all feature urban interstate segments estimated to carry more than 250,000 vehicles per day.

The Volume/Capacity (V/C) ratio compares the actual traffic on a highway link to the link's carrying capacity and is a primary measure of highway traffic congestion. A V/C ratio between 0.71 and 0.95 indicates that traffic volume is between 71 percent and 95 percent of capacity -- a warning signal to transportation planners. A highway with a ratio greater than 0.95 is considered at capacity or congested. Reno's (1988, 380) data reveal that between 1981 and 1986, interstates became the most congested of all urban highway types (Table 2.2). By 1986, 24.9 percent of all U.S. urban interstates were congested. During the same period, the number of congested rural interstates also increased but at a much lower level. Only 8.8 percent of all U.S. rural interstates operate near or at capacity with only 2.9 percent of all rural routes categorized as congested.

In retrospect, given congestion as the lone criterion, it is obvious that urban interstates were under-built while their rural counterparts were over-built. A number of factors contributed to the inaccurate traffic forecasts leading to capacity imbalances. Between 1950 and 1970, the U.S. population increased by 34.3 percent. Automobile registrations skyrocketed (Figure 2.1). The U.S. driving public willingly redefined its travel behavior and took to the roads in private automobiles. The suburbs of major cities grew and changes in the nation's economy created new demands on its transportation system.

The industrial and military "needs" of the 1950s defined a set of criteria around which the system was built. Designers wanted

TABLE 2.1
U.S. Highway Mileage and Vehicle Miles Traveled

	Mileage				
	1950	1960	1970	1980	1990
Total	3,312,975	3,545,693	3,730,082	3,856,858	3,877,540
IHS	0	10,440	30,027	40,253	42,436
Rural	0	9,250	24,051	29,865	31,583
Urban	0	1,190	5,976	10,388	10,853
Other	3,312,975	3,535,253	3,700,055	3,816,605	3,835,104
Rural	2,990,036	3,106,875	3,145,361	3,203,761	3,089,241
Urban	322,939	428,378	554,694	612,844	745,863

	Vehicle Miles Traveled (Millions)				
	1950	1960	1970	1980	1990
Total	458,246	718,845	1,109,724	1,527,295	2,147,501
IHS	0	No data	161,048	296,326	478,977
Rural	0	No data	79,516	135,084	200,573
Urban	0	No data	81,532	161,242	278,404
Other	458,246	718,845	948,676	1,230,969	1,668,524
Rural	239,998	387,260	459,956	536,946	669,836
Urban	218,248	331,585	488,720	694,023	998,688

Source: U.S. Department of Transportation, Federal Highway Administration 1985c, 1990a, 1991b.

to foster the family car as the main mode of personal transportation and to ease the surface shipment of heavy weapons. In fact, increasing commerce and boosting U.S. defense capabilities are goals that have characterized the federal government's thinking

TABLE 2.2
Volume/Capacity Ratios by Highway Type and Area

Highway Type and Area	Percent V/C Ratio from 0.71 to 0.95		Percent V/C Ratio Greater than 0.95	
	1981	1986	1981	1986
Urban				
Interstate	14.5	20.4	16.6	24.9
Other expressway	11.4	14.7	15.8	18.8
Other principal arterial	16.4	16.8	18.7	18.0
Minor arterial	11.7	9.8	11.3	11.2
Collector	4.5	4.1	4.5	3.4
Rural				
Interstate	2.3	5.9	0.6	2.9
Other principal arterial	2.7	1.8	1.6	1.1
Minor arterial	2.0	1.1	1.0	0.6
Major collector	0.6	0.2	0.7	0.1
Minor collector	0.1	0.1	0.2	0.0

Source: Reno 1988.

about transportation for 200 years (American Trucking Associations, Inc. 1967). The direct result is that in 1990, U.S. citizens purchased more than 114,262,125,000 gallons of gasoline, registered more than 188,655,462 vehicles, and drove more than 2 trillion miles (U.S. Department of Transportation, Federal Highway Administration 1991b, 6, 17).

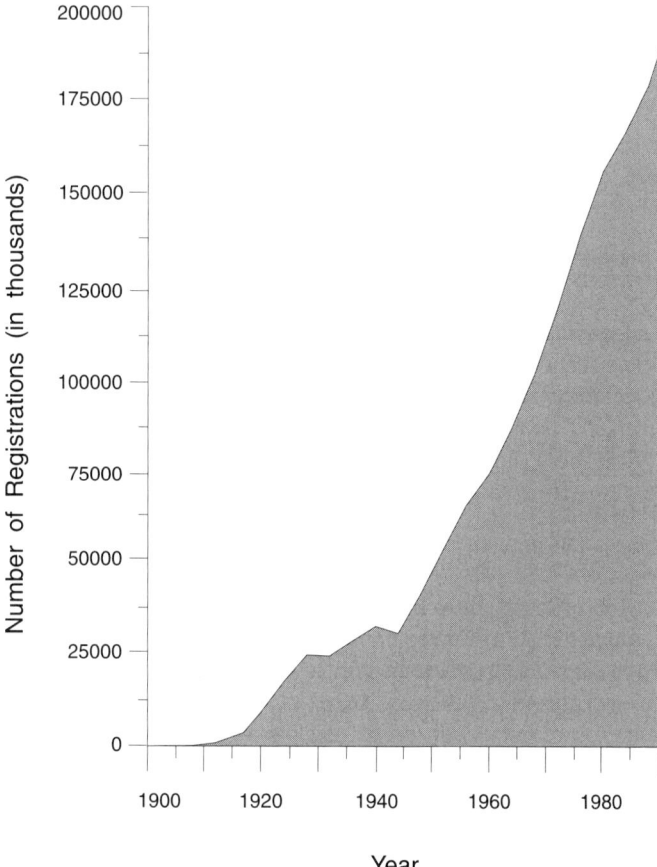

Figure 2.1. U.S. automobile registrations, 1900-1990. Sources: U.S. Department of Commerce, Bureau of Public Roads 1957b; U.S. Department of Transportation, Federal Highway Administration 1985c, 1991b.

Highway Safety

Despite the increase in population and vehicle miles traveled on U.S. highways during the development of the IHS, highway fatality rates measured by population, licensed drivers, registered vehicles, and vehicle miles traveled decreased between 1966 and 1990 (Figure 2.2) The estimated number of U.S. highway traffic fatalities for 1992 (around 39,500) was expected to be the lowest annual figure in 30 years (Traffic 1992).

There remains, however, considerable geographic variation across the states as illustrated by changes in the number of fatal traffic crashes between 1989 and 1990 (Figure 2.3). Experiencing the steepest declines in traffic fatality rates were the District of Columbia, Vermont, Rhode Island, New Hampshire, Massachusetts, Nebraska, Pennsylvania, Arkansas, and Iowa (U.S. Department of Transportation, National Highway Traffic Safety Administration 1991, 39). Compared to these states is another group where fatal crash rates increased: North Dakota, Hawaii, Montana, Alaska, Delaware, and Kentucky (U.S. Department of Transportation, National Highway Traffic Safety Administration 1991, 39).

Improvements in traffic fatality statistics since the 1960s are attributed to enhanced alcohol abuse awareness programs, increased law enforcement, widespread seat belt usage, and newer

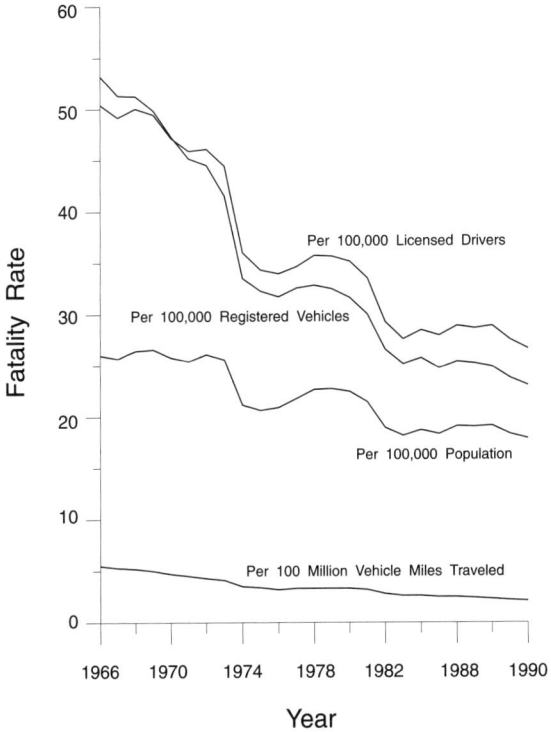

Figure 2.2. U.S. traffic fatality rates, 1966-1990. Source: U.S. Department of Transportation, National Highway Traffic Safety Administration 1991.

30 National Impacts of the IHS

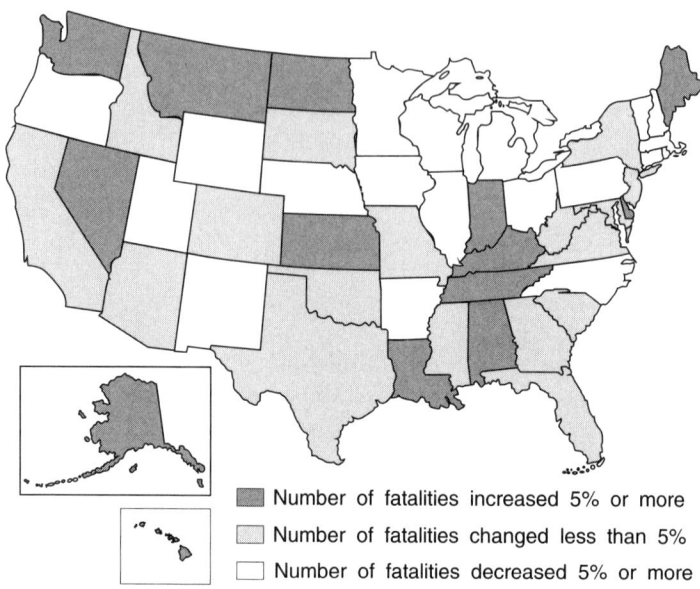

Figure 2.3. Statewide patterns of change in the number of fatal motor vehicle crashes from 1989 to 1990. Source: U.S. Department of Transportation, National Highway Traffic Safety Administration 1991.

and better roads, including the IHS. Figure 2.4 compares the number of fatal traffic crashes across ten roadway function classes. Interstates are the scene for fewer fatal crashes than occur on almost all other major roadways and, because of the high number of vehicle miles traveled on the interstates, have a lower rate of fatal crashes per million VMT than other roadway classes. Even critics of the IHS agree that it has resulted in significant reductions in traffic accidents. In a comparative analysis of 7,000 highway miles located across 39 states, the Federal Highway Administration (1970c, 4) found that accident rates were lowered in both rural and urban areas and across accident types. Property damage accidents decreased 38 percent in rural and 48 percent in urban areas. Personal injury accidents decreased 39 percent in rural and 37 percent in urban areas, and fatal accidents decreased 43 percent in rural and 15 percent in urban areas.

Another more subtle and difficult to measure negative user impact of the IHS involves crime. Development of the IHS has given rise to "freeway crimes" including commuter shoot-outs in Los Angeles, objects thrown through windshields from overpasses in Detroit, and large caliber sniper-fire aimed at IHS travelers near Jacksonville. Among law enforcement agency personnel, I-75 is known as "cocaine lane" (Drug mules...1993, 9). The use of IHS routes as major drug distribution arteries necessarily impacts system users.

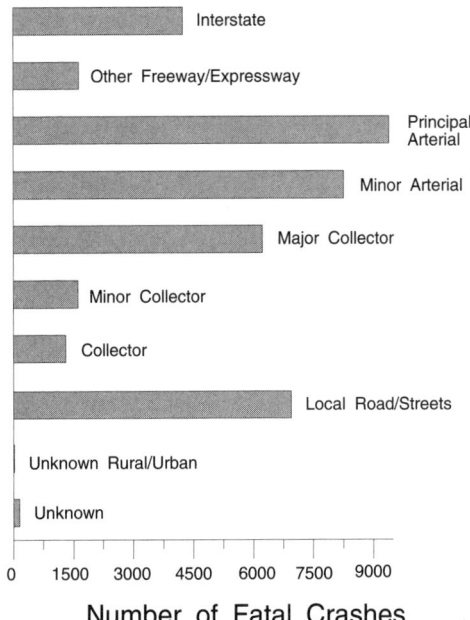

Figure 2.4. Fatal crashes by highway function class, 1990. Source: U.S. Department of Transportation, National Highway Traffic Safety Administration 1991).

The Reduction of Functional Distance

As early as 1844 Ralph Waldo Emerson recognized a fascinating geographic process that he labeled the *"abolition of distance"* (quoted in Wolfe 1963, 117). Emerson was simply noting the way trains and steamships were making travel easier and quicker. He was referring to what others have since called "*the shrinking earth*" -- a reduction in the importance of physical distance. Although the absolute distance between points remains constant, enhanced transportation technologies bring places closer together in terms of the relative time and cost of interaction between them. Therefore, it is the **time** between places, as opposed to the absolute distance, that is abolished (or minimized) by transportation advances (Wolfe 1963).

As a network, the IHS has greatly reduced functional distances between locations in North America. Places that were relatively remote and isolated in 1950 are more accessible today because of the network. Accessibility refers to the relative location of a network's nodes or, in the case of the IHS, the points that collectively form the landscape of the U.S. Even if a city or town is bypassed by the network and spatially separated from it by the lack of a local interchange, it may still be more accessible than it was prior to the system's construction.

Taaffe and Gauthier (1973) compared accessibility via the IHS with that via the national railroad network. A particular concern was the spatial structure of the two networks and the associated spatial relationships among U.S. cities in each system. Taaffe and Gauthier calculated an accessibility score for each U.S. city by counting the number of its links with other cities. A city's score reflects its relative location to other cities. By comparing accessibility scores, the levels of accessibility enjoyed by each city in each transportation system were evaluated. Indianapolis was the most accessible U.S. city on the railroad network, followed by Louisville and Cincinnati (Table 2.3). The most accessible cities on the IHS were Louisville, Nashville, and Indianapolis.

When analyzed in their entirety, the two lists reveal a regional shift in relative accessibility toward the Southeast (Figure 2.5). Large portions of Ohio, Indiana, and Illinois are replaced in the national "core" -- those places characterized by the very highest levels of accessibility -- by places in Kentucky, Tennessee, and Georgia. Detroit, Milwaukee, and Grand Rapids are among the cities falling from the top 25 while Bristol, Charleston, and Greenville are among those moving onto the highway list. Taaffe and Gauthier (1973, 148) explain any similarities between the two sets of rankings as a reflection of the trunk line planning approach brought to the IHS and the tendency of transportation planners to *"replicate and thereby reinforce an existing spatial organization."* In this case, the pattern of the national railroad network was somewhat duplicated by that of the IHS.

The importance of the "trunk line" planning approach is considered by Patrick (1991) who documents the existence of "communication bundles" crisscrossing the U.S. A communica-

TABLE 2.3
National Railroad Network and the IHS City Accessibility Rankings

Railroads		Interstate Highways	
Rank	City	Rank	City
1.	Indianapolis	1.	Louisville
2.	Louisville	2.	Nashville
3.	Cincinnati	3.	Indianapolis
4.	Dayton	4.	St. Louis
5.	St. Louis	5.	Lexington
6.	Chicago	6.	Columbus
7.	Columbus	7.	Dayton
8.	Lexington	8.	Knoxville
9.	Toledo	9.	Cincinnati
10.	Memphis	10.	Chattanooga
11.	Spartanburg	11.	Chicago
12.	Birmingham	12.	Atlanta
13.	Detroit	13.	Toledo
14.	Chattanooga	14.	Memphis
15.	Nashville	15.	Bristol
16.	Little Rock	16.	Akron
17.	Knoxville	17.	Cleveland
18.	Milwaukee	18.	Birmingham
19.	Kansas City	19.	Muskegon
20.	Moline	20.	Charleston
21.	Atlanta	21.	Youngstown
22.	Akron	22.	Moline
23.	Cleveland	23.	Greenville
24.	Grand Rapids	24.	Kansas City
25.	Muskegon	25.	Pittsburgh

Source: Taaffe and Gauthier 1973.

34 National Impacts of the IHS

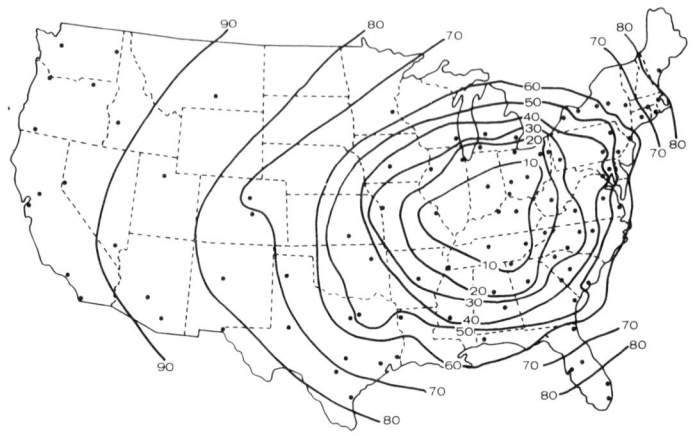

Figure 2.5. IHS accessibility isolines. Source: Taaffe and Gauthier 1973.

tion bundle is a set of spatially contiguous links from different networks that might include a railroad line, an interstate highway, a utility line, or a fiber optics cable. Their presence together indicates a high-order path of connectivity between nodes that facilitates the varied interactions between places. Regional dissimilarities can be smoothed by evolving networks as places and people increasingly interact. Increased accessibility leads to greater spatial interaction, and that interaction has measurable impacts on the regions surrounding the network. Heightened accessibility translates into higher volume flows of both passengers, goods, and information.

The impacts of the IHS on increasing accessibility and spatial interaction across the U.S. are well documented. On a national scale, the IHS significantly reduced the average travel time between places. The Federal Highway Administration claimed a system-wide reduction factor of ten percent within the first decade of operation (U.S. Department of Transportation, Federal Highway Administration 1970c, 3).

Social Impacts of the System

The impacts of the development of the IHS on people are many, affecting both users and non-users. These impacts can also be positive or negative, sometimes pitting the interests of users against those of non-users. Social impact assessment of transportation systems frequently involves evaluation of changes in accessibility and in personal mobility. Personal mobility is different from accessibility in that it addresses one's ability to travel and not his or her location relative to a set of possible destinations. Accessibility is a function of distance regardless of how distance may be measured, while personal mobility depends on a host of social and personal factors including income, physical ability, and age. The transportation literature is rich in case studies on the impact of interstate segments and interchanges on individuals and families.

Personal Mobility

Residents of the United States enjoy the highest level of personal mobility in the world partly because of the IHS. The data on vehicle registrations graphed in Figure 2.1 shows the increasing number of road vehicles registered. The number of private and commercial automobiles in the U.S. per capita has grown to 0.57, more than one vehicle for every two people (Figure 2.6) (U.S. Department of Transportation, Federal Highway Administration 1991b, 16). Roughly 87 percent of U.S. households have at least one automobile and 53 percent have two or more (Reno 1988, 373). U.S. adults licensed to drive average roughly 30 miles of personal travel per

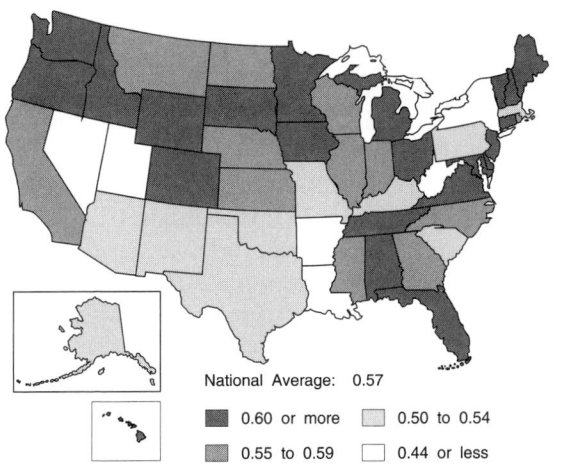

Figure 2.6. Per capita automobile ownership, 1990. Source: U.S. Department of Transportation, Federal Highway Administration 1990b.

day (Reno 1988, 373). The total number of miles traveled in privately owned automobiles increased from 775,940 million in 1969 to 1,409,202 million in 1990, an increase of 82 percent (U.S. Department of Transportation, Federal Highway Administration 1991b, 211).

Over the same period, vehicle operating costs were reduced. Design attributes of the system such as lower grades, reduced traffic friction, and removal of stop signs and lights generally lower operating costs (U.S. Department of Transportation, Federal Highway Administration 1970c). Along with individual drivers, firms and consumers benefit from lower operating costs. Cost reductions in travel have made extended commuting possible and reduced travel time to other activity sites, too. For example, the national park system is now accessible to millions of potential visitors who were more remote from these locations prior to the IHS.

Automobile Dependency

Today's citizens are more dependent on personal automobile travel than earlier generations. This dependency relationship is in part a result of the IHS. Eighty-seven percent of individual trips to work take place in privately owned automobiles (74 percent as drivers and 13 percent as passengers) (Reno 1988, 376). Home-to-work trips are the most common type of travel, accounting for 26 percent of all trips and 33 percent of all miles traveled. The average distance in miles of the journey to work decreased from 9.4 in 1969 to 9.2 in 1977 to 8.6 in 1983 (U.S. Department of Transportation, Federal Highway Administration 1991b, 215-217). After this slight decline, the average distance lengthened in the late 1980s, reaching 10.9 miles by 1990. While many factors contribute to this pattern, interstates certainly play a role. Interestingly, drivers seem to be increasing their speeds to offset this increase in trip distance (Reno 1988). Despite this behavior, the length of time spent on the journey to work is steadily increasing (American Association of State Highway and Transportation Officials 1987).

Travelers have fewer competitive travel options because the U.S. intercity passenger transportation system has been impacted by the network's presence. The rail system suffered most of the passenger losses. The IHS along with the growing commercial aviation industry reduced the market share held by intercity rail services from 6.5 percent in 1950 to 0.7 percent in 1990 (Eno 1991,

47). Mass transit operators find it difficult to compete with convenient and highly subsidized urban interstates.

A higher share of personal and family incomes, even among non-users, is devoted to transportation, as network construction and maintenance costs are collected through non-fuel taxes. Roughly 21 percent of U.S. household expenditures are for transportation (Reno 1988, 376). Among the social costs of transportation levied against all members of our society, drivers and non-drivers alike, are those of police protection, emergency service, traffic monitoring and control, planning, and the environmental and social costs of automobility. There is clearly a positive relationship between income, automobile ownership, average annual transportation expenditures and, in essence, personal mobility (Table 2.4). The growing demand for access and mobility among the populace has been tied to the development of open roads.

Forced Relocation and Local External Effects on Neighborhoods

Perhaps the most severe IHS non-user social impact is forced relocation. Thousands of eminent domain proceedings have been used to uproot residents living in the path of proposed highway construction. Once in place, superhighways can serve as barriers, dividing and functionally separating neighborhoods and subdividing farms, as city streets and country lanes alike are permanently severed by dissecting routes.

The remaining residents are drawn into the realm of influence of the highways, interchanges, and traffic. The influence is not limited to the IHS routes *per se* but diffused into surrounding neighborhoods. Outside of roughly 125 feet of a freeway, derived convenience outweighs nuisance (Thiel 1962). In spite of evidence that urban freeways are congested, traffic analysts agree that the IHS has reduced congestion on nearby streets and roads (see Table 2.2), in some cases up to 50 percent (U.S. Department of Transportation, Federal Highway Administration 1970c).

The negative local impacts of the IHS are not limited to the nation's cities but affect suburban, exurban, and rural areas. Many of the preexisting corridors through the rural U.S. have declined in importance. Among the best examples are Route 66, the Dixie Highway, and the National Highway. IHS critics argue that much of the development attributed to the system was relocated from these earlier corridors. Development of the IHS across the U.S. caused the country to reflect on its social landscapes and in many

TABLE 2.4.
U.S. Automobile Ownership and Transportation Expenditures
by Income

	First Quintile	Second Quintile	Third Quintile	Fourth Quintile	Fifth Quintile	All
Automobile Ownership						
Average Number of Vehicles	1.0	1.5	1.9	2.4	3.0	1.9
Percent with at least 1 vehicle	58	84	92	95	97	85
Vehicles per Person 18 or Older	0.7	0.9	1.1	1.1	1.3	1.0
Average Annual Automobile and Related Expenditures						
Vehicles	677	1,133	1,902	2,540	4,056	2,063
Gasoline and Oil	538	781	1,011	1,314	1,578	1,045
Vehicle Expenses	509	762	1,126	1,549	2,258	1,241
Other	136	168	222	229	568	265
Total	$1,860	$2,844	$4,261	$5,632	$8,459	$4,614
Income Spent on Transportation						
Percentage	16.9	20.1	22.2	21.7	20.0	20.5

Source: Reno 1988.

cases consciously choose between preservation and deterioration. The choice, however, was often made in drafting rooms and engineering offices instead of city halls and county courthouses. The negative impacts on the local landscape from the IHS are partly responsible for the introduction of social impact assessment, now commonly used across the U.S. in land use planning.

Economic Impacts

IHS impacts on the U.S. economy have been debated since the system's inception. The economic impacts of transportation innovation often pit community against community and region against region. Here too, the literature offers many case studies of IHS segments and interchanges and how they economically impact places (The University of Iowa Public Policy Center 1990).

Local Economic Stimulus

Highway construction directly and indirectly brings federal tax dollars into local communities. After construction, interstate highways can function as catalysts for local economic development and employment, especially in rural areas where the economic impact of an IHS route and interchange may be greatest (see Chapter Five and Moon 1987a). According to one of the most basic geographic tenets, increased accessibility to previously isolated places enhances their value. From the urban perspective, increased freight shipments by truck have been shown to increase the market size of larger urban centers (U.S. Department of Transportation, Federal Highway Administration 1970c). At the same time, rural and urban employees benefit from expanded job opportunities that were previously less accessible and employers gain a larger pool of potential employees (see Chapter Four). And, from a retail trade perspective, rural and suburban residents often gain new shopping opportunities as local and regional businesses experience increased accessibility and visibility. As a result of all of these interrelated, system-driven phenomena, rising property values along with higher employment rates and access to higher paying jobs serve to collectively increase local tax rolls.

Negative Local Economic Impacts

Residents immediately adjacent to the IHS often suffer a reduction in the value of their properties. Consequently, property tax rolls can be diminished by the property acquisition necessary to build the highway segments. From a land use perspective, more downtown space is devoted to parking as more commuters drive greater distances to CBD employment (see Chapter Four). Among businessed displaced by the IHS, the rate of survival for small businesses was less than for larger businesses (U.S. Department of

Transportation, Federal Highway Administration 1976d). Even worse, the entire functions of some businesses were eliminated or redefined. A case in point is the motor court of the pre-interstate period. Finally, from the perspective of a small town, IHS routes that functionally serve as bypasses have often decreased Main Street's viability as a retail and service location.

Positive Regional and National Economic Impacts

Reduced travel time and increased accessibility have resulted in the easier and less costly movement of goods across the U.S. In the first decade of system construction, the highway ton miles of freight shipped on the IHS increased by more than 60 percent (U.S. Department of Transportation, Federal Highway Administration 1970c). In many cases, distribution costs are lower by truck on the IHS.

Truck size has increased with IHS development resulting in lower shipping costs for certain products, especially those that are bulky and relatively lightweight. Some IHS routes now permit the use of double and even triple trailer combinations. Increasing truck size and accessibility over the IHS have advanced intermodal transportation. Several railroad companies now use advanced truck-train combinations in addition to the containerized shells that are lifted from truck to train to ship. The new truck-trains involve the attachment of axle and wheel assemblies underneath trailers. These assemblies allow the trailer to function as a railroad flat car.

The flow of venture capital into previously underdeveloped areas increases with improved accessibility, as discussed in Chapter Four. At the same time, commercial and industrial firms have more options when evaluating expansion, location, or relocation, as described in greater detail in Chapters Three, Four, and Five. A case in point is the relatively rural Southeast of the pre-IHS period before the Sunbelt was opened for development by the IHS. Attractive southern locations for commercial outlets, industrial plant sitings, and retiree residential developments became much more accessible from the northeast and midwest and the region experienced significant new development. Tourism and recreation opportunities drew industries catering to travelers. Within these historically underdeveloped regions, strategically located interchanges were often the focal points of new urban and suburban growth areas (see Chapter Six and Moon 1989) .

TABLE 2.5
Modal Distribution of U.S. Products Shipped, 1950-1990

Mode	1950	1960	1970	1980	1990
Rail	46.7	36.1	31.1	28.7	26.6
Road	26.1	32.7	36.2	36.2	40.8
Water	17.9	18.2	17.1	17.8	16.1
Air	0.0	0.0	0.1	0.1	0.1
Pipeline	9.3	13.0	15.6	17.3	16.4

Source: Eno Transportation Foundation, Inc. 1991. Reprinted with permission of Eno Transportation Foundation, Inc.

Negative Regional and National Economic Impacts

The U.S. transportation system has been reconfigured by the network's presence with the railroad system suffering the greatest impact. An examination of freight shipments by mode since the 1950s illustrates these changes (Table 2.5). The share of the nation's shipped products carried by rail fell from almost 50 percent to almost 25 percent between 1950 and 1990. Not apparent from these figures is the spatial dimension of truck versus rail shipments. Trucks provided a higher level of competition at the local and regional levels than at the national level, particularly during the early development of the IHS (U.S. Department of Transportation, Federal Highway Administration 1970c). Rail shipment of many goods remains competitive at the larger scale.

While decentralization of economic activity was occurring in metropolitan areas as individuals, families, commercial establishments, institutions, and industrial firms moved from the city center to the metropolitan periphery, there was also a decentralization of economic activity from the urban manufacturing core region. The Northeast and Midwest have experienced the highest out-migration of people and firms. The Southeast and Southwest were primary destination regions and urban centers in these areas experienced explosive growth (Baerwald 1978).

Environmental Impacts

The impact of the IHS on the environment is not very well understood or documented. Surprisingly to some, cited impacts are both positive and negative. However, in virtually every case of interstate construction/renovation involving environmental questions, environmentalists are lined up against those promoting economic development.

Early research reported by the FHWA indicated that freeway and interchange construction could provide benefits in the control of soil erosion (U.S. Department of Transportation, Federal Highway Administration 1970c). Among the new landscapes features created by IHS construction are borrow pits. Borrow pits are the depleted quarries left behind following construction of interchanges in particular. Adjacent to the network around most of the U.S., borrow pits provide some recreational opportunities but have more often been left as artificial wetlands. Drainage patterns are altered during construction. In the early years of IHS construction, natural wetlands were filled and/or drained with little or no impact assessment. From a larger perspective, perhaps the greatest nationwide environmental impact generated by the IHS came from the conflicts surrounding its construction. The development of the IHS spurred the adoption of the Environmental Impact Statement (EIS) as a requirement in the transportation and land use planning process across the U.S.

While some of the most ardent IHS supporters might contend the point, most observers agree that IHS development has produced landscapes that are not aesthetically pleasing. This is in spite of the FHWA's assertions that urban freeways are "visually pleasing," in some cases *"works of art,"* and that they provide *"a pleasing aesthetic experience"* (U.S. Department of Transportation, Federal Highway Administration 1970c, 31). Noise, air, water, and light pollution accompany the IHS through neighborhoods and across farms. Undeveloped land is altered and/or paved often resulting in increased soil erosion (Hamilton 1988), contradicting earlier FHWA claims.

An indirect environmental consequence of the IHS is the increased accessibility to previously isolated places. Entire ecosystems are redefined or destroyed (for example, parts of the national park system overrun by visitors). Sensitive plant and animal communities are threatened not only during construction but thereafter by traffic and the pollution it generates. Research

TABLE 2.6
Top Ten State Traffic Fatality Rates

Rank	State	Fatality Rate per 100,000 Vehicle Miles Traveled
1.	Nevada	3.4
2.	West Virginia	3.1
3.	New Mexico	3.1
4.	Mississippi	3.1
5.	Arkansas	2.9
6.	South Carolina	2.8
7.	Alabama	2.6
8.	Florida	2.6
9.	Montana	2.5
10.	Louisiana	2.5

Source: U.S. Department of Transportation, National Highway Traffic Safety Administration 1991.

indicates higher lead concentrations in some animal populations sampled along the IHS (Irwin, *et al.* 1981).

Given the negative environmental impacts experienced in areas of IHS construction, there have been some renowned cases of environmental impact mitigation in the planning and construction of IHS segments. Among these is the case of Glenwood Canyon, Colorado, where construction was delayed for more than 20 years while a route less harmful to the environment than the route originally planned could be designed and "fitted" into the sensitive canyon located 160 miles west of Denver near the Utah border (Chamberlain and Sorrentino 1991). This 12-mile segment, the last leg of I-70, cost more than $490 million, features two 4,000-foot tunnels, and 40 bridges. A number of exceptions to the IHS design and construction standards were made for this segment

(Crowds... 1992, 10). *The Federal-Aid Highway Act of 1976* allows deviations from federally mandated standards in cases like Glenwood.

Conclusion

IHS analysts have only a general level of understanding of cumulative U.S. traffic levels on the IHS, and we do not know specifically where that traffic occurs. There has also been a dearth of research on the impact of the IHS on the physical environment. Much more information is available on the social and economic impacts of the IHS (Table 2.6), particularly in metropolitan areas where population densities are high and localized social impacts have been great. The next two chapters examine in greater detail the social and economic impacts of IHS development in U.S. metropolitan areas and regions.

3

THE IHS AND METROPOLITAN AREAS

> *One thing you can set down as sure is that cities are doomed. (Henry Ford quoted in Patton 1986, 97)*
>
> *We shall solve the city problem by leaving the city. (Henry Ford quoted in Flink 1990, 139)*
>
> *[The Hollywood Freeway is] the longest parking lot in the world. (Edmund Brown quoted in Smerk 1965, 75)*
>
> *That bloated eight-lane serpentine monstrosity that slithers down ... is set and ready to strike the totally unprepared city with a crippling injection of 40,000 more cars per day this fall. (Michael Stedman quoted in Rose 1990, 101)*
>
> *The raised stub of the partially built Embarcadero [Freeway] was left standing forlorn and purposeless, unless it was to be seen as a war memorial to freeway fighters. (Dunn 1981, 123)*

The focus of this chapter is the impact of IHS development on metropolitan areas. Many issues at the national scale discussed in Chapter Two are amplified in metropolitan areas where dynamic populations and the land they control react to reduced transportation costs and changing patterns of accessibility. The network capacity and volume questions raised earlier are especially critical in metropolitan areas where the system has been overloaded almost from the start. The related issues of highway safety and reduced functional distance are more pronounced in metropolitan areas and nowhere is the system's impact on the economy, the environment, and society greater than in urban America.

Locational Conflict in Urban Areas

The early construction of interstate routes inside U.S. cities caused an outbreak of conflict over the system's intent, design, location, and even existence. Through what became known as the "Freeway Revolt," development of the IHS met with unexpected urban opposition (Dunn 1981, 123). The engineers that brought the IHS to the people really believed that they were addressing one public and, to make matters worse, assumed that the public was "presold" on the IHS concept and the plan to run segments of the system through U.S. cities (Kelley 1971, 92). While those opposed to construction were routinely accused of representing an *"organized campaign against the automobile"* (Kelley 1971, 92), they lacked a focus -- except halting the construction of urban interstates. According to Kelley (1971), the many groups loosely organized in opposition to the system were separated by geography, time, and narrowness of immediate purpose. Despite the fact that there was no coordinated nationwide effort to block the IHS, the impact of these groups in individual communities was felt at the highest levels of government and industry.

Ex post facto public involvement in IHS planning predetermined and set the stage for confrontation. Supporters and critics of the IHS disagree over the extent of the organized opposition. Most agree that in at least two dozen U.S. cities there was enough conflict to temporarily, if not permanently, halt construction (Kelley 1971). It is generally accepted that the conflict's greatest impact was on the urban planning process itself, for the planning process was changed by community level opposition regardless of how local. The list of cities experiencing "superhighway contests" and the mileage in question was presented to Congress in 1969 by the Federal Highway Administration (Table 3.1). Among these cities, San Francisco, Washington, and Boston stand out as classic examples of what can happen when a transportation plan meets with strong public opposition.

San Francisco

In San Francisco, opposition to the idea of widespread urban freeways through the city peaked in 1959 (Kelley 1971). The impetus for this opposition was the planned Embarcadero Freeway across historic and scenic Market Street. Early freeway plans featured six elevated traffic lanes but because of anticipated traffic

TABLE 3.1
Locational Conflicts Over Urban Highways in 1969

City	Length of Contested Segments (miles)	Nature of Conflict
Atlanta	4.0	neighborhood impact, displacements
Baltimore	5.5	displacements, costs, historic sites
Boston	3.1	displacements, route justification
Charleston	2.0	displacements, esthetics
Cleveland	8.8	parkland, historic sites, religious interests
Detroit	7.4	neighborhood impact, tax loss
Indianapolis	6.5	displacement, esthetics
Memphis	3.7	parkland
Nashville	4.4	neighborhood impact, access
Newark	7.2	neighborhood impact, displacements
New Orleans	3.1	historic sites
New York	25.8	displacements
Philadelphia	7.3	displacements, esthetics, relocation, housing
Pittsburgh	3.4	displacements
San Francisco	17.4	displacements, parkland, esthetics
Washington	24.2	displacements, costs, parkland, esthetics

Source: Kelley 1971.

volumes the plans were redrawn to include eight. Caught between federal and state highway officials and an uneasy public, San Francisco officials proposed an underground location for the freeway but this was rejected outright. In what Dunn (1981, 123) refers to as *"one of the wildest San Francisco board meetings on record,"* the city's Board of Supervisors decided to halt the entire project. The Embarcadero Freeway was left hanging virtually in mid-air throughout this period of activism that evolved into a new era for Bay-area politics (Figure 3.1). The halted project has since grown to be a symbol not only of Bay area activism and resistance but of urban preservation at the local level in the face of federal intervention.

Through powers granted California cities via state law, San Franciscans were able to resist a highway that they did not want. The energy aroused by the proposed Embarcadero Freeway was then rechanneled behind its successor, the Bay Area Rapid Transit System (BART). The city became the first and only to refuse federal highway funds outright and instead voted to impose a $792 million bond issue to initiate construction of BART (Kelley 1971, 95). Refusing to accept $280 million in highway funds, Bay area citizens and politicians agreed to accept an economic burden exceeding one

Figure 3.1. The Embarcadero Freeway. Source: California Transportation Photo Department, Undated.

billion dollars. However, other U.S. cities lacked the freedom or will to make such a choice.

Washington, D.C.

In the Nation's capital, opposition to federally mandated superhighway construction peaked ten years later. Dominant among the many aspects of this "freeway revolt" was the whole issue of statehood for the District of Columbia. This controversial issue, in combination with the city's housing of the national government, set the stage for a unique battle between high-level planners/administrators and citizens. To District of Columbia freeway fighters, the Three Sisters Bridge across the Potomac River between the city and suburban northern Virginia was comparable as a symbol to the Embarcadero Freeway in San Francisco.

Transportation in and around Washington, D.C. has long been the focus of debate at every level of government (Hebert 1972). Diffusion of the automobile simply accelerated an evolving crisis, pushing it to the forefront of late 1960s local and national politics. Kelley (1971, 109), once a Federal Highway Administration public affairs director, writes that the city is *"already crisscrossed with more freeway miles per capita and per square mile than any other in the U.S., including symbolic Los Angeles."* Hebert (1972) confirms this notion of the automobile's role in the city by citing its number one ranking among U.S. cities in terms of per capita automobile ownership. By the late 1960s, Washington, D.C. was overrun on a daily basis with more than 800,000 vehicles (Hebert 1972, 172). The vast majority of this number consisted of privately owned cars -- more than 4,000 per square mile (Hebert 1972, 172). To address the mounting urban congestion brought on by automobiles, Congress had earlier created the National Capital Transportation Agency to develop a multi-modal solution. The agency's 1962 report called for 50 additional miles of superhighways including the Three Sisters Bridge. Despite the fact that the city lacked the political freedoms and mechanisms of other units protected by state charters, citizen reaction to the 1968 adoption of the regional transportation plan was swift and harsh (Hebert 1972).

Another interesting aspect of this case study involves the conflicting desires of city residents and suburbanites from surrounding communities. The transportation plan in question was designed to promote regional accessibility and therefore was welcomed in suburban Maryland and Virginia. However, citizens of the District saw it as a means to further congestion, and, inevitably to greater out-migration. According to Hebert (1972, 172), they cited the plan's impressive costs and potential negative impacts including:

[1] a $65 million annual price tag for 8 to 10 years;
[2] construction through several of the city's "finest" neighborhoods;
[3] a loss of 180 acres of residential property;
[4] 15,000 displaced residents;
[5] the loss of 225 acres of commercial property;
[6] 245 lost acres of parks, monuments, and other government owned land;
[7] a loss of 41 acres of institutional property; and
[8] an annual loss from the local tax base of $6 million.

But because state involvement was impossible and the city's government was so well connected to the federal power base, early resistance to the plan was ignored. Even after a newly elected city council voted to halt implementation of the plan, federal level politicians and planners refused to back down. Not only did they refuse to adhere to local wishes, Congressional action in 1968 led to the withholding of federal transit funds until the freeway portion of the plan was underway (Hebert 1972).

Eventually, federal court action on behalf of local citizens stopped parts of the plan from being carried out. While some freeway miles were constructed, the Three Sisters Bridge was not and its absence signifies a victory in the battle for home rule in the city. Because this face-to-face confrontation between local citizens and the federal government occurred on the government's doorstep, it played a prominent role in the development of grassroots activism in the United States.

Boston

Perhaps the most renowned case of citizen opposition to urban interstates occurred in Boston. Citizens there are credited with "metropolitanizing" the fight against freeways as groups with vastly divergent interests from across the Boston metropolitan area joined forces against a common enemy -- Boston's planned Inner Belt and its related network of arterials (Lupo, Colcord, and Fowler 1971, 16). An inner belt to bypass the central city had been proposed since 1948 but, prior to passage of *The Federal-Aid Highway Act of 1956*, had remained unfunded (Lupo, Colcord, and Fowler 1971). With passage of this federal legislation came the 90/10 funding split that made the project possible and plans for its construction were brought down from the shelf. According to a 1969 plan, Boston's inner belt was to be an eight-lane freeway through some of the city's most well-known neighborhoods -- Roxbury, the Fenway, Brookline, Cambridge, Somerville, and Charlestown (Weiner 1986).

Boston's inner belt is a classic example of a project born from top-down planning by aggressive administrators. The controversial federal Secretary of Transportation, perhaps the nation's leading superhighway advocate at the time, was a former director of the Massachusetts Public Works Department and former governor. In addition, the incumbent Massachusetts governor throughout the controversy had also been director of the state's Public Works Department and was a strong inner belt proponent. In fact, he went on record saying that, *"This road is the key to the entire Massachusetts system. It has to be built"* (Dunn 1981, 124). However, a coalition of representatives from neighborhood and civil rights groups and local churches joined with community leaders and a "blue ribbon panel of intellectuals" from area universities to oppose the project (Dunn 1981, 124). Among the most vocal of the organizations against the inner belt were the Greater Boston Committee on the Transportation Crisis and the Save Our Cities Committee. Together these groups forced Mayor Kevin White in 1969 to reverse his earlier support of highway construction. When Boston's mayor was finally convinced to oppose the entire project, discussion of an inner belt was taken to a higher level of public debate and eventually removed from further consideration.

Never before had such a regionally representative and broad-based set of groups opposed a public works endeavor. The scale of the confrontations matched the scale of the projects. Out of the Boston experience came means for assessing the regional impacts of transportation plans. The Boston Transportation Planning Review (BTPR) is still recognized as the model for social and environmental impact analysis. In this approach, an interdisciplinary team of regional leaders and residents examine various transportation alternatives and assess their effects on the community. In the Boston case, the plan took 18 months to complete and not only doomed the inner belt but established a new *modus operandi* for regional transportation planning across the country.

The "Freeway Revolt" impacted all of transportation in the U.S. by educating the general public in terms of the system and the many issues tied to it and by legitimizing the protest against *"a highway-dominated [U.S.] transportation system"* (Dunn 1981, 124). While responses ranged from noisy public hearings to protests and threats, citizen involvement in transportation planning was initiated. According to Dunn (1981, 124) the opposition created an *"image of the masses rising in wrath against the greedy, intensive road builders."* After this image was firmly in place among the urban citizenry, federal policy shifted more funds to other modes of transportation resulting in the transfer of some control and decision making to localities. The origin of BART in San Francisco and METRO in Washington, D.C. can be traced to the intent and nature of the federal highway planning approach adopted for implementation of the IHS. Later legislation, specifically *The Federal-Aid Highway Acts of 1970* and *1973*, enacted many of the community-oriented changes to the planning process that had emerged during the prior decade.

Central Business District (CBD) Impacts

Central cities had good reason to contest the IHS because the development of urban expressways has had a dramatic effect on their economic and social viability. It would be difficult to overstate the impact of the automobile on American cities. McKelvey (1973, 112) writes that *"no other physical agent could match the automobile in its influence on the development of the metropolis in America."* Urban IHS routes accelerated and expanded the impact of the automobile on the location of residential, business, industrial, and retail activity.

Robertson (1980) cites increased personal mobility as the critical component behind urban population decentralization. With increased accessibility to the suburbs via urban interstates, city dwellers experienced more freedom and more locational options in residential choice. Many residents no longer had to make the trade-off between low density living and accessibility to downtown -- they could now have both. Consequently, residential growth in the suburbs increased (Robertson 1980). The IHS was a key factor in this process. Table 3.2 illustrates the change in inner city population as a percentage of metropolitan area population. For these cities, the IHS had its greatest impact on residential shift between 1960 and 1980.

TABLE 3.2
Inner City Population as a Percentage of
Metropolitan Area Population

City	1960	1970	1980	Change in Inner City Population, 1960-1980
Boston	30.6	26.7	23.5	-23.2%
Chicago	57.1	48.2	42.3	-25.9%
Denver	53.1	41.9	30.9	-41.8%
Detroit	44.4	36.0	29.8	-32.9%
Houston	35.2	26.3	16.5	-53.1%
Los Angeles	41.1	39.9	39.7	-3.4%
New York	55.1	48.7	45.4	-17.6%
Phoenix	11.0	7.0	3.7	-66.4%
San Francisco	27.9	23.0	20.9	-25.1%
Washington, D.C.	36.8	26.4	21.4	-41.8%

Source: Newman and Kenworthy 1989.

The exodus of selected central city populations supported by enhanced transportation also exerted a great influence over downtown manufacturing (Robertson 1980). Because of decentralization and greater accessibility, industrial activity was no longer tied to the central city. Firms could now relocate to the periphery and still have access to a diverse and abundant urban labor force. According to Robertson (1980), non-central city manufacturing locations have a number of advantages over urban locales, including more available space for horizontal assembly and processing plants, lower property costs, the prestige of a suburban address, and agglomeration benefits in industrial parks. Urban interstates facilitate the industrial community's tapping of these advantages. In a comprehensive study of manufacturing growth in 212 U.S.

urban areas, Wheat (1969) substantiated a strong but disparate IHS influence in various regions of the country. There was a clear, positive interstate impact east of the Mississippi and in the far west. These regions are *"characterized by dense population and uneven terrain -- regions where freeways offer relatively substantial time savings"* (Wheat 1969, 24). Firms in these areas benefitted from the presence of the IHS. Wheat found a positive correlation between proximity to the highway and an industry's economic viability. Successful firms migrated toward the IHS and the suburbanizing labor followed.

The extent of this migratory process is illustrated in Table 3.3. The data indicate inner city employment as a share of metropolitan area employment for ten U.S. cities. In every case except one -- Los Angeles -- the cities experienced a decline in inner city employment relative to metropolitan employment. There is a strong correlation

TABLE 3.3
Inner City Employment as a Percentage of Metropolitan Area Employment

City	1960	1970	1980	Change in Inner City Employment, 1960-1980
Boston	46.2	40.5	34.6	-25.1%
Chicago	66.5	52.4	44.9	-32.5%
Denver	61.0	57.7	49.7	-18.5%
Detroit	56.6	39.8	29.6	-47.7%
Houston	na	na	41.1	na
Los Angeles	38.7	42.8	43.3	+11.9%
New York	54.1	48.5	41.9	-22.6%
Phoenix	23.3	18.3	10.6	-54.5%
San Francisco	43.4	37.9	34.4	-20.7%
Washington, D.C.	59.2	46.7	32.5	-45.1%

Source: Newman and Kenworthy 1989.

between the residential change noted in Table 3.2 and the employment change cited in Table 3.3.

Finally, Robertson (1980) notes the marked influence of decentralization on downtown retailing. The development of suburban shopping centers, malls, and strip centers coincided with the population shift. Among the main advantages for suburban retailers over their downtown competition were accessibility, less traffic congestion, and spatial division of markets. In addition, suburban retailers often offered more convenient business hours, easy and inexpensive parking, perceived safety, a more appealing arrangement of stores, and a more aesthetically pleasing atmosphere. Robertson (1980) also recognizes the lack of organization among downtown merchants, land owners, and public officials. The cumulative result of these suburban advantages and downtown disorganization has been the steady flow of retail trade to the suburbs from the old CBD. Urban freeways serve as the arteries along which this flow occurs and suburban interchanges are the prime receptors. The rapid and large scale suburban construction of retail outlets stands in sharp contrast to the empty storefronts and shops that characterize many downtown areas.

Another interesting aspect of urban decentralization and suburbanization is the amount of available parking in CBDs. In spite of the fact that inner city populations are decreasing along with the level of downtown employment, more CBD parking is demanded by suburban commuters. Since most commuters travel alone on the journey to work, there is an inordinate demand for convenient parking in the CBD. Table 3.4 documents the growth of CBD parking as a land use in ten U.S. cities. Across urban America, historical preservationists have fought the paving of downtowns to facilitate parking. Downtown merchants and administrators, on the other hand, have urged the demolition of low-density and/or vacant structures to accommodate the mounting flow from suburbia.

Social Impacts

The decentralization of urban populations, manufacturing, and retailing plus the expansion of downtown parking redefines the lives of those left behind. Critics of the IHS and the neighborhood destruction it entailed in U.S. cities claim a blatant racial bias in the location of its routes (Black leaders... 1992). Interstate planners frequently sought out low property cost paths for their

TABLE 3.4
Parking Spaces in the CBD

City	1960	1970	1980	Change in Number of Spaces, 1960-1980
Boston	29,300	53,800	70,200	+139.6%
Chicago	26,781	35,109	35,374	+32.1%
Denver	31,800	38,890	49,919	+57.0%
Detroit	36,600	44,500	52,400	+43.2%
Houston	28,293	34,928	64,194	+126.9%
Los Angeles	64,900	68,826	80,074	+23.4%
New York	na	na	144,926	na
Phoenix	14,174	22,000	26,772	+88.9%
San Francisco	29,300	36,400	39,665	+35.4%
Washington, D.C.	49,029	65,061	70,943	+44.7%

Source: Newman and Kenworthy 1989.

projects, deflecting them to neighborhoods where property values were depressed. A 1976 FHWA policy statement addressing the discrimination issue is straightforward and clear-cut.

> *Locating highways where they will minimize disruption to residential neighborhoods emphasizes the stability of the neighborhood. Neighborhoods that may be particularly vulnerable to freeway disruption and therefore to be avoided are high density pedestrian-dependent neighborhoods with few autos available and strong racial or ethnic ties (U.S. Department of Transportation, Federal Highway Administration 1976d, xii).*

Opponents argue that this statement and the professed protectionist attitude of the freeway planners came about after the fact, if at all. As evidence, an analysis of neighborhood disruption in Baltimore due to urban interstate construction finds clear discrimination (Christensen and Jackson 1969, 2). Of the 3,800 families (15,000 people) displaced by the urban interstates in question, 80 percent were non-white and fewer than 40 percent were homeowners. When IHS routes ran through cities, large numbers of people and businesses were displaced. Worsening this displacement, in the eyes of the authors of the study, was that expressways were routed through the *"least desirable sections of the city, and those who are displaced are the poor, the aged, and those who are least able to take care of themselves, and there is little likelihood that many of them will use the expressway that displaces them"* (Christensen and Jackson 1969, 1).

Compounding these problems, minority communities fail to enjoy many of the benefits of freeways found in non-minority communities. In a comparative study of 17 cities, Steptoe and Thornton (1986) contrast the freeway impacts in minority and non-minority communities:

[1] minority businesses experience great hardships in relocating and often close when forced to move;
[2] freeways through minority communities fail to attract the level of economic development found in the other study areas;
[3] there are minimal differences in land use and housing quality in minority neighborhoods before and after freeway construction;
[4] vacant land remains available around many interchanges in minority neighborhoods; and,
[5] the property value increase attributed to freeway proximity in non-minority communities is absent in minority neighborhoods.

Researchers have not adequately addressed the placement of freeways through minority, aged, and/or low income communities. Nor have they thoroughly documented the impact (or lack thereof) of urban interstates on sensitive and disadvantaged neighborhood economies. Consequently, there is an extreme shortage of research on this topic. But those knowledgeable of cities such as Detroit, Baltimore, and Cincinnati recognize the freeway's propensity for "finding" these possible paths of least resistance through the city. And, we know that the negative impacts associated with the IHS led some cities to "trade in" billions of federal construction dollars for other projects as allowed in the *The Federal-Aid Highway Act of 1973* (Bloch and Crowell 1984).

Beltway Development

Urban beltways reign as predominant symbols of the effect of the IHS on metropolitan morphology. Beltways can be defined as limited-access expressways partially or completely encircling larger cities. While some isolated cases predate the network, most beltways were constructed as parts of the IHS. *The Federal-Aid Highway Act of 1956* authorized construction of 2,300 miles of urban beltways (5.8 percent of the total authorized mileage) primarily designed to bypass urban areas (U.S. Congress 1956a, 1956b; Payne-Maxie, *et al.* 1980, 5). Their impact on urban decentralization and its counterpart, suburbanization, has been dramatic. In the late 1970s, the U.S. Department of Transportation and the Department of Housing and Urban Development (HUD) jointly funded a national-level assessment of beltways and their impact on urban morphology, land use, and nearby residents. The extensive study centered on a comparative analysis of 27 cities with beltways and 27 without.

The study (Payne-Maxie, *et al.* 1980) found that in most cities, beltways were built on open land and integrated into the existing roadway system often based on radials or a grid pattern. They also noted that few federal guidelines figured into beltway construction and that local officials generally approached the building of beltways from a technical perspective. For all practical purposes, the social and economic effects of beltway design were not taken into consideration.

The following beltway characteristics are cited by the authors. Complete loops did not generate more impacts or traffic than partial loops. Beltway location and interchange spacing were the most important design features, affecting land use, urban development, and, at a larger scale, metropolitan development. Increased accessibility through closely spaced interchanges and frontage roads attracted development around beltways but the distance between a CBD and its beltway was not a significant factor in development. Generally, highly accessible interchanges were more attractive locations for commercial and industrial development. In terms of the interaction between beltways and the cities that they surround, the consultants noted that beltways can increase traffic on intersecting streets as well as on the radial highways that often function as links between key interchanges and the CBD. Beltways also played important regional roles in shaping trips not oriented to the downtown, the flow of through traffic, and in serving as cross-town connectors between suburban residential clusters.

Among the 27 cities surrounded by beltways analyzed in this study were Atlanta (I-285), Baltimore (I-695), Columbus (I-270), Louisville (I-264), Minneapolis (I-494), Omaha (I-680), Raleigh (I-40), and San Antonio (I-410). Figure 3.2 illustrates the evolutionary nature of beltways and the pattern of nearby regional shopping center development. Based on the 54 city analysis, the consultants drew the following important conclusions. A beltway can *"increase development opportunities, support prevailing urbanization patterns, and facilitate compact development,"* yet they find that, in general, beltways do not improve a metropolitan area's competitive advantage (Payne-Maxie, *et al.* 1980, 10-11). This finding directly addresses the notion of "metropolitan dominance" via *"expanding metropolitan communities"* purported earlier by Berry (1960, 113). Apparently, certain interchanges affect the location, timing, size, and initial success of regional shopping centers, and accessibility and visibility attract industrial and office park developers to certain corridors and interchanges. In terms of residential growth, beltways are particularly apt to spur the development of multi-family residences. Counter to these peripheral processes, CBDs surrounded by beltways feature lower gains in retail sales and employment as well as higher office vacancy rates. The consultants summarize that most of the initial economic and land use impacts of beltways result from the transfer of activities outward from the CBD as opposed to new development from other regions. They also note that beltways negatively effect the disadvantaged and low income residents of central cities.

In a detailed examination of beltway development around Minneapolis-St. Paul, Baerwald (1978) models the pattern of development around a seven-mile segment of I-494. He identifies four stages of corridor development leading to what he terms a "new downtown." The first stage consists of post-World War II residential development followed by a second stage of industrial diversification and commercial expansion. A third stage consists of speculative development fostered by a strong economy and results

60 The IHS and Metropolitan Areas

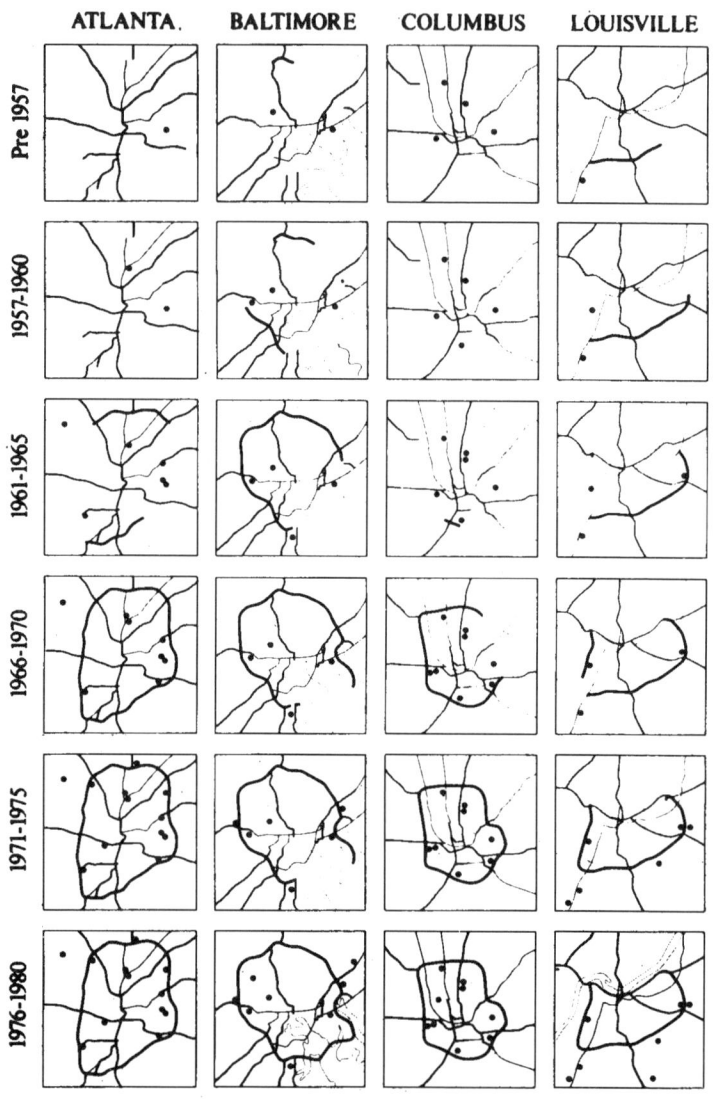

Figure 3.2. Evolution of eight beltways. Source: Payne-Maxie, *et al.* 1980.

Beltway Development 61

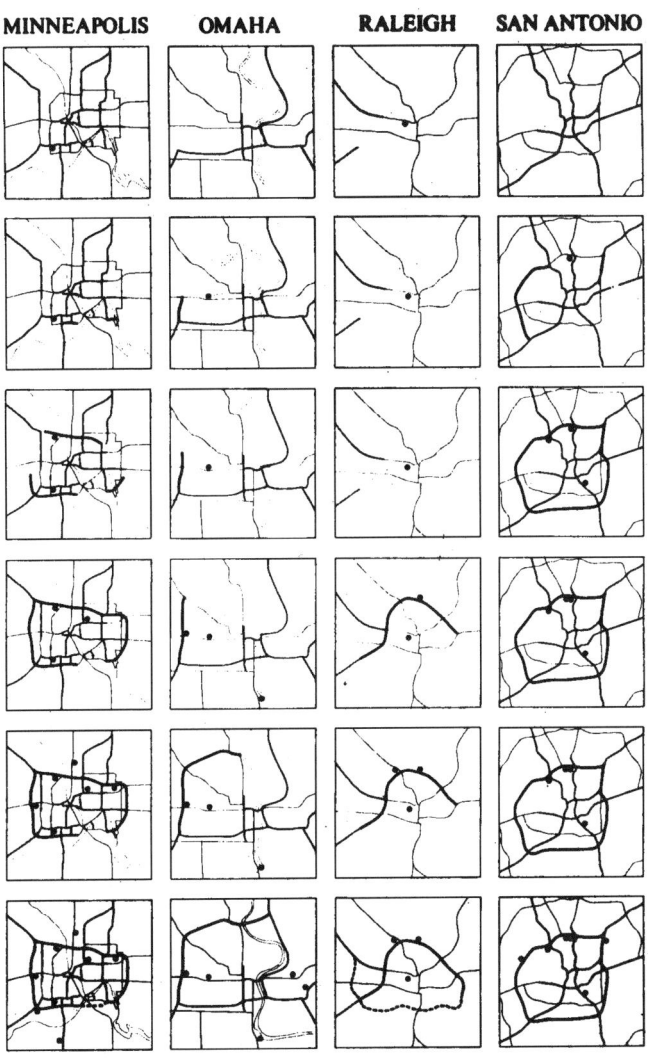

● Regional Shopping Center

in dramatically increased property values. The fourth and final stage of corridor development involves redevelopment wherein the function, value, and overall nature of the retail sector is redefined. Figure 3.3 maps the development of the I-494 corridor over time (Baerwald 1978).

Interestingly, the corridor at the center of Baerwald's study is now the site of the Mall of America, the world's largest mall. The 78-acre facility has 2.5 million square feet of retail floor space in a massive retail/dining/entertainment complex largely funded by Japanese investors (Teachers Insurance and Annuity Association 1992). Mall operators plan to hire up to 10,000 employees to accommodate an expected 40 million patrons annually.

In a follow-up study, Baerwald (1982) distinguishes between corridor and cluster development patterns around beltways. Unlike the linear development corridor that he documented in the initial study, development clusters occur around key nodes (interchanges). Certain land users are attracted to clusters rather than corridors resulting in significantly different land use patterns. Comparison goods stores, high-rent residences, medical facilities, and direct-access public services tend to agglomerate as node-oriented cluster development. On the other hand, convenience stores, supportive public services, lower-rent residences, automobile dealers, and industrial firms/warehouses tend to disperse along development corridors. A development cluster can sometimes occur within a development corridor.

Erickson and Gentry (1985) expand on this work by documenting the growth of "suburban nucleations" from 1957-1980. They define suburban nucleations as development clusters around key freeway interchanges. They are sprawled around larger urban areas functioning as "urban subcenters" or "mini-cities". In this model of nodal development, firms realize the benefits of agglomeration as they draw each other to attractive suburban locales. Here, Robertson's (1980) summary of the positive features of suburban location or relocation come into play. According to Erickson and Gentry, an interchange functioning as a receptor for this interest should connect the freeway and a preexisting main arterial. At these key junctures, "interactive nodes" operate as volatile suburban land markets similar to those found at certain CBD cores. Figure 3.4 illustrates the evolution of a suburban nucleation. In summary, the authors identify lower transportation costs and land rents as factors behind development of suburban nucleations. The work of Baerwald and Erickson and Gentry

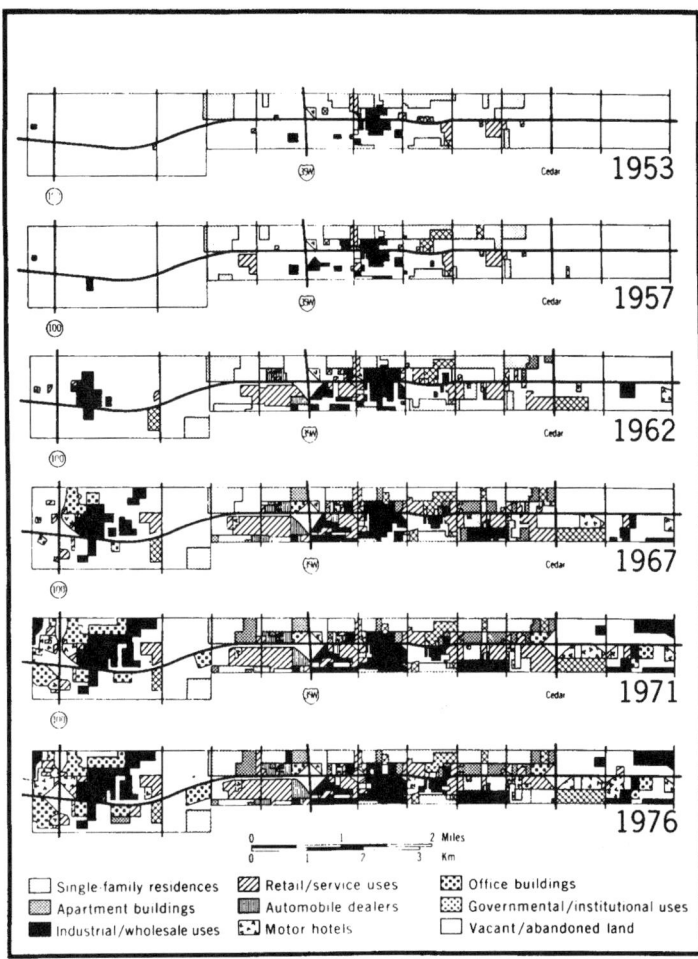

Figure 3.3. Land use in the I-494 corridor. Source: Baerwald 1978. Copyright The American Geographical Society. Reprinted with permission.

serves as the basis for a number of later studies on office location (Smith and Selwood 1983) and suburban gridlock (Cervero 1986).

One City's Experience

Los Angeles is often cited as a city heavily influenced by automobile transportation in terms of both its form and function. Its freeway system, comprised almost entirely of IHS routes, is the

64 *The IHS and Metropolitan Areas*

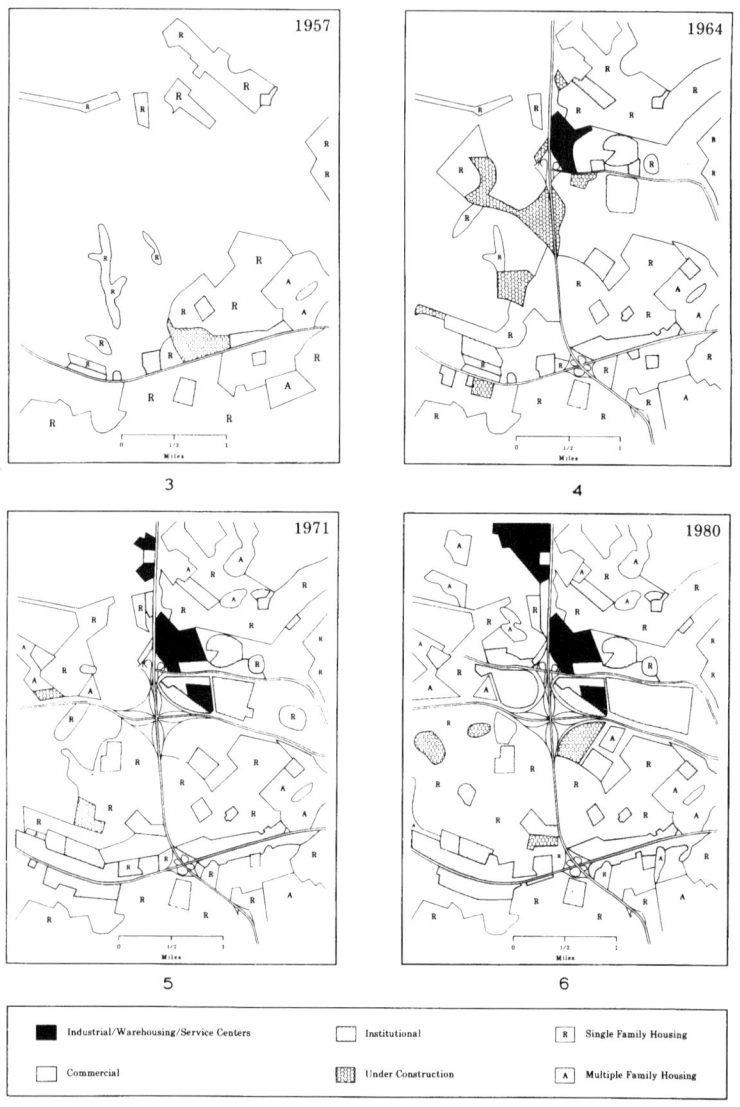

Figure 3.4. Evolution of a suburban nucleation. Source: Erickson and Gentry 1985. Copyright The American Geographical Society. Reprinted with permission.

most famous in the world (Organization for Economic Cooperation and Development 1988). The city is the heart of an urbanized area whose population of roughly ten million is scattered over an area of nearly 2,000 square miles (Lowry 1988, 296). The urbanized area features 17 major activity centers, each covering more than 300 acres with over 5,000 jobs (Gordon, *et al.* quoted in Lowry 1988, 296). Perhaps more indicative of the urbanized area's function as an employment generator, less than 20 percent of the total number of jobs in the area are within one of these 17 centers. The City of Los Angeles is the largest center, encompassing more than 6,700 acres and providing more than 373,000 jobs. According to recent estimates by the Southern California Association of Governments, only 5.1 percent of the urbanized area's population lives in areas where residential density is 30 or more persons per acre (Lowry 1988, 297). More than 80 percent of the population resides in areas with less than 20 persons per acre. These data paint a graphic picture of a sprawling metropolitan area born of the freeway era.

Personal mobility is very important in southern California and nowhere more than in Los Angeles. Data on a number of personal mobility measures illustrate the increasing role of the automobile there during the freeway era (Table 3.5). The Los Angeles urbanized area has more than 13,000 miles of arterials including roughly 2,000 freeway miles and another 60,000 to 70,000 miles of city streets (Organization for Economic Cooperation and Development 1988, 79). Within Los Angeles County, freeways comprise only one percent of total highway mileage but carry 55 percent of the total VMT (Organization for Economic Cooperation and Development 1988, 79).

Government spending on Los Angeles highways virtually doubled during the 1970s, an era noted for population increases, two-hour traffic jams, and widespread road construction. By 1980, the average number of daily trips in the urbanized area reached nearly 39 million (Organization for Economic Cooperation and Development 1988, 83). Roughly 53 percent of these trips involved the journey to work but only about 1 million trips (2.6 percent of all trips) were made via public transit. A very high 40 percent of all VMT (88.4 million of 221 million) occurs during peak travel periods. The urbanized area has, in at least one year, consumed more than 6 billion gallons of gasoline and diesel fuel (Organization for Economic Cooperation and Development 1988, 83).

In Los Angeles County, over half of the days in some years have violated national ozone standards with concentrations over 0.20

TABLE 3.5
Dimensions of Personal Mobility in the Los Angeles Area

Mobility Variable	1960	1970	1980
Passenger Cars/1,000 People	459.1	521.2	541.5
Total Vehicles/1,000 People	513.0	615.0	666.7
Parking Spaces/1,000 CBD Workers	372.8	534.8	523.6
Mean Work Trip Length	19.5	24.7	24.7
Percent Work Trips by Private Vehicle	85.2	89.2	88.0
Per Capita Car Miles Traveled	11,907	12,662	14,522
Per Capita VMT	12,535	13,615	16,135
Per Capita Car Occupant Miles	17,980	19,120	22,364
Per Capita Vehicle Occupant Miles	18,927	20,559	24,848
Per Capita Road Length	38.8	34.5	30.4
Vehicles Per Road Mile	181.3	220.2	254.2
Car VMT in Millions Per Road Mile	2.4	2.6	3.2
Vehicle VMT in Millions Per Road Mile	2.6	2.8	3.6
Total VMT Per Car	25,937	24,293	26,818
Total VMT Per Vehicle	24,434	22,137	24,200

Source: Newman and Kenworthy 1989.

parts per million (Organization for Economic Cooperation and Development 1988, 83). At the same time, noise levels as high as 85 to 90 decibels have been measured in the area. Until very recently, local leaders took few legitimate steps to develop a viable, non-highway, regional mass transit system since the infamous demise of the Red Lines of one half century ago. But with the advent of a mass transit system in this region recognized as the bastion of the private automobile, the travel behavior of its residents may change, and such a change could have widespread implications. According to Lowry (1988, 97), Los Angeles serves as a *"bellwether for future urban growth"* in an area of the country that is undergoing rapid growth and where *"Los Angeles-style urbanization is the rule rather than the exception."*

Conclusion

After decades of effort, pressure from the private sector finally contributed to the Congressional decision to address a national highway network of predominantly freeways -- the IHS. But the result of that pressure was met by stiff opposition, especially in urban areas. Heymann (1965) attributes this popular uprising against urban expressways, that began in the late 1950s, to the lack of a national transportation plan -- a true focus. He points out that the logical first question in transportation planning should go to its purpose but in fact the question is hardly ever asked. In addition to asking *"how much and where"* he charges planners and citizens to ask "what for" (Heymann 1965, 18). That question was asked by millions during debate over the future of urban interstates but goes unanswered today because of a broad nationwide, divergent, and often conflicting set of transportation objectives. Ironically, those varied objectives led to a substantive change in U.S. transportation policy and perspective -- development of the regional transportation planning perspective.

In spite of thirty years of regional planning in the U.S., it can be argued that the evolution of the post-IHS metropolitan area has been anything but planned. Many CBDs exist as shells of what they were before the proliferation of urban expressways and beltways. Their industries are gone and suburban malls, shopping centers, and strip malls represent the relocation of downtown retailing to the periphery. The social and safety impacts of the IHS on urban America have been dramatic and not equitably spread among its residents (Table 3.6). These secondary and tertiary effects of the interstate network were probably beyond the thoughts of early freeway fighters when they took on the federal transportation planning bureaucracy but their concerns for cities physically destroyed by the IHS were not without merit. The early and often verbalized fears of the residents of San Francisco, Washington, D.C., and Boston are realized in today's Los Angeles where proliferation of the family car stifles the city's transportation system and clouds its atmosphere with daily regularity.

TABLE 3.6
Top Ten Urban Traffic Fatality Rates

Rank	City	Fatality Rate per 100,000 Population
1.	Orlando, Florida	27.3
2.	Tampa, Florida	27.1
3.	San Bernardino, California	26.2
4.	Miami, Florida	23.4
5.	Dayton, Ohio	23.1
6.	Altanta, Georgia	19.5
7.	Montgomery, Alabama	19.2
8.	Kansas City, Missouri	18.4
9.	Hialeah, Florida	17.6
10.	Nashville, Tennessee	17.4

Source: U.S. Department of Transportation, National Highway Traffic Safety Administration 1991.

4

THE IHS AND REGIONAL DEVELOPMENT

Roads, from the beginning of time, were built for commerce.
(Roland Rice quoted in Rose 1990, 59)

If the interstate system does not turn out to be an extraordinary benefit, it should definitely be classified as man's most appallingly wasteful project. (Smerk 1965, 131)

In his critique of government spending in transportation, Nelson (1973) listed four parts of our national transportation system built more for political reasons than to address regional transportation needs: marginal inland waterways; many county roads; some airports; and the rural interstate system. In each case, he cited the "enormous waste" of tax dollars in the construction and upkeep of these components. In addition to these examples of "over investment" in transportation, Nelson (1973, 258) pointed to federal *"regulatory policy and uneconomic pricing of freight services"* (Nelson 1973, 258). The government's focus on highway transportation has been a prime source of widespread regional and national changes in U.S. transportation. Among these are changes in the modal division of passenger and freight traffic, the location and style of urban living, the form and function of cities, the roles played by small rural towns and villages, and industrial relocation (Nelson 1973, 230). Many of these changes generate large scale regional impacts -- effects that ripple across sizable areas of the U.S. This chapter examines IHS impacts on the development of regions.

The Network and Regional Redistribution

Whether they have been an over investment or not, federal expenditures on transportation have increased rapidly. Table 4.1 docu-

TABLE 4.1
Government Expenditures for Transport Service and Facilities
by Mode and Year, 1960 - 1990 (in millions)

Mode and Fund Source	1960	1965	1970	1975	1980	1985	1990
Airways - Total	429	652	965	1,652	2,135	2,263	4,323
Federal	429	652	965	1,652	2,135	2,263	4,323
State/Local	-	-	-	-	-	-	-
Airports - Total	421	510	1,080	1,790	3,157	4,623	7,418
Federal	79	95	111	342	656	879	1,263
State/Local	342	415	969	1,448	2,501	3,744	6,155
Highways - Total	10,160	13,456	19,502	27,207	39,188	55,715	70,100
Federal	2,753	4,137	5,181	7,180	12,036	15,092	15,000
State/Local	7,407	9,319	14,321	20,027	27,152	40,623	55,100
Harbor/River - Total	524	667	820	1,268	2,324	2,684	3,097
Federal	287	391	376	526	1,156	1,189	1,097
State/Local	237	276	444	742	1,168	1,495	2,000
Railroads - Total	-	-	40	342	2,088	2,361	2,438
Federal			40	299	1,405	1,407	1,188
State/Local			0	43	683	954	1,250
Transit - Total	-	64	166	3,017	7,201	8,316	9,136
Federal		51	133	1,589	3,541	2,216	1,573
State/Local		13	33	1,428	3,660	6,100	7,563
All - Total	11,534	15,349	22,573	35,276	56,093	75,962	96,512
Federal	3,548	5,326	6,806	11,588	20,929	23,046	24,444
State/Local	7,896	10,023	15,767	23,688	35,164	52,916	72,068

Source: Eno Transportation Foundation, Inc. 1991. Reprinted with permission of Ento Transportation Foundation, Inc.

ments the enormous outlays for transportation at the federal, state, and local levels in the U.S. These figures reflect a total increase of spending on transportation nearing 750 percent from 1960 to 1990 (Eno Transportation Foundation, Inc. 1991, 73). In addition, the investments are far from equally distributed across the six dominant modes. Of the $96,512 million invested by the U.S. government in transport facilities and services between 1960 and 1990, highways garnered $70,100 million. This figure, equalling 72.6 percent of the total, reveals the dominance of the automobile, truck, and bus during the freeway era.

Analysis of regional development impacts often involves comparisons of transport innovation costs versus benefits. The cost-benefit aspects of regional highway systems are difficult to gauge because roads are built to meet a variety of different objectives. Justifications might range from the reduction of traffic congestion, to spurring economic development, to meeting some latent transportation demand. Highway funding may also be appropriated to meet some political and/or social goal such as the redistribution of regional wealth. In fact, one impact of the IHS has been regional wealth redistribution.

Through studying more than thirty years of contributions into and withdrawals from the Highway Trust Fund, a clear pattern of regional transfer can be established. Figure 4.1 depicts the receipt/payment ratios for all 50 states over the period 1957 to 1990. If a state has a ratio of less than 1.00, its population has contributed more dollars into the Highway Trust Fund than the state has received. Such states are known as "donors" and are located mostly east of the Mississippi. The "recipients" of their generosity are generally located in the western U.S. With a ratio of 7.67, Alaska benefits from an incredibly high return on its population's payments into the trust fund. If judged on its ability to redistribute wealth from the eastern to the western states, the IHS probably is

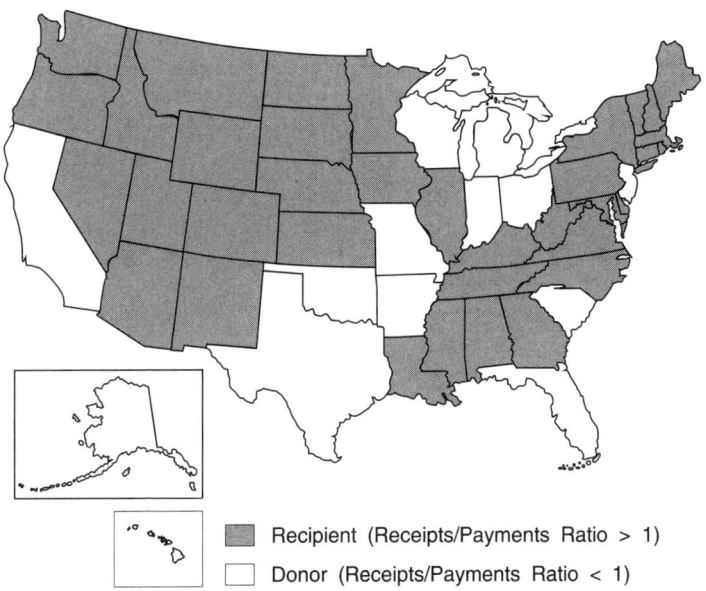

Figure 4.1. Federal Highway Trust Fund Receipt/Payment Ratios. Source: U.S. Department of Transportation, Federal Highway Administration 1991b.

matched only by the defense industry. The average receipt/payment ratio of states east of the Mississippi is 1.27 compared to the ratio for western states of 1.81.

Transport Costs Versus Benefits

Economists disagree over the network's benefits relative to its costs. Friedlaender (1965, 64) went so far as to write that:

> *The traffic in rural areas traversed by the Interstate System is too small to support an entire network with the planned capital intensity, and a somewhat smaller urban Interstate System in conjunction with mass transit would probably have been closer to the optimal investment for the urban component of the Interstate System.*

Others write of the enormous shares of the GNP "wasted" on transportation. Nelson (1973) believed that the developmental advantages of highway investments are offset by their cost. In his view, an inefficient federal investment policy favoring highway transportation has resulted in the decline of railroads and created exorbitant costs for society in the ensuing modal misallocation of traffic. Highway have *"produced the lion's share of the most discouraging of the social costs from transport. These include excessive noise, health damaging smog, and a high accident toll of fatalities, personal injuries, and property damage each year"* (Nelson 1973, 252).

In a detailed 1990 examination of state level transportation expenditures and potential returns on that investment, the University of Iowa's Public Policy Center in conjunction with the Midwest Transportation Center found a mixed bag in terms of derived benefits. After comparing the expense of road building against a reduction in transportation costs for users, the Center reported the following conclusions (The University of Iowa Public Policy Center 1990, 2-3):

[1] a state can promote economic development by undertaking highway investments if and only if they are efficient;

[2] a road project generates benefits only to the extent that it lowers transportation costs, broadly defined to include safety and environmental impacts;

[3] even though a road project may be seen as an economic development tool, it must be justified on the basis of its transportation benefits alone;

[4] when a road project is evaluated as a means of attracting a particular business to a certain location, it must be compared to other means of attracting the business, such as direct financial assistance in the form of a cash subsidy, a low interest loan, or a tax abatement;

[5] when assessing the benefits of road projects, it is necessary to recognize that building or improving a particular stretch of road may reduce the benefits derived from existing highways; and,

[6] road investment is sometimes seen as a means of redistributing income, wealth and/or prosperity within a state, but it is a poor tool for doing so.

In contrast to these findings, Prentiss (1962), writing in an earlier era, considered highway building an employment stimulus. He cited 1958 figures indicating 94,138 hours of construction site employment from each million dollars spent on the system. Off-site employment was thought to be greater than or equal to on-site employment. Prentiss estimates the multiplier of IHS construction at 1:4, with every dollar spent on the network stimulating the spending of four additional dollars. More recent studies argue for lower ratios, but generally continue to support that network impacts are positive.

On the other hand, most analysts early and later in the system's development agree on the illogic of using highway investment as an anti-recession tool. Prentiss (1962) summarized the general opinion of the time by citing the unique advantages that highway building offers as an economic activity capable of easing a national recession. Federal financing of the national highway program, however, *"does not, and should not, contemplate periodic spurts in the highway program to counter cyclical downturns in the economy"* (Prentiss 1962, 356).

Wilson, *et al.* (1985) found that estimating the regional effects of highway construction is difficult because highway development occurs in three distinct stages. In the initial stage, scattered, disconnected highway nodes and links evolve into an actual network with very limited regional economic impact. A second stage of highway development generates regional economic development as a surrounding area responds to the influx of external capital. Interestingly, the main regional response to highway construction is further road building. Finally, a third stage ensues wherein regional development slows in response to highway construction. However, this third and final stage is typified by a marked increase in the personal mobility of the region's residents. The timing of a region's transition through the three stages varies depending on a variety of local and regional conditions. Wilson, *et al.* (1985) conclude that improvements in our transportation infrastructure provide the "opportunity" for economic development, but because of the uniqueness of U.S. regions, each might react to the opportunity at different times with different results.

The Network and Regional Migration

Geographers agree that development of the IHS has had a profound effect in redistributing the U.S. population. In a nationwide analysis of interstate population migration between 1935 and 1970, Clayton (1977) found that both the process and structure of migration are changing at two different levels -- local and national. At the local level, migration occurs among contiguous states while at the national level it takes place in definite patterns across many states. At both levels, migration is increasing as the distance between places becomes less important. Increasingly sophisticated transportation and communication technologies have had measurable impacts on the ease of movement across the United States, with California and Florida emerging as important migration "basins" in the 1970s (Clayton 1977, 180).

Fuguitt and Beale (1976) examined population change in nonmetropolitan U.S. cities and towns between 1950 and 1970. Total nonmetropolitan population growth in counties with interstate highways was found to be double that of nonmetropolitan counties without interstate highways. In counties with cities of 10,000 people or more, the effect was reduced, suggesting that the network was playing a more important role in rural areas. Among these counties and inside incorporated places, the role of freeways was apparently less important (Table 4.2).

In an analysis covering 1950 to 1975, Briggs (1980) detailed similar findings. Briggs studied every nonmetropolitan county in

TABLE 4.2
IHS Impact on Percent Population Change

Percent Change from 1950 to 1960

	All Places		Incorporated Places		Unincorporated Places	
Initial Size of Largest Place in County	On IHS	Off IHS	On IHS	Off IHS	On IHS	Off IHS
Less than 2,500	-1%	-6%	9%	3%	-4%	-9%
2,500 - 9,999	4%	-1%	12%	10%	-1%	-8%
More than 10,000	18%	15%	16%	19%	17%	10%
All Places	12%	5%	15%	14%	10%	-2%

Percent Change from 1960 to 1970

Less than 2,500	8%	1%	10%	6%	6%	1%
2,500 - 9,999	10%	3%	10%	8%	10%	-1%
More than 10,000	14%	12%	10%	11%	19%	12%
All Places	13%	7%	10%	10%	15%	4%

Source: Fuguitt and Beale 1976.

the continental U.S. Counties with freeways exhibited higher migration rates than those located off the network, and, in areas already undergoing high migration, counties with open freeways registered even higher migration rates than those without open freeways. In addition, counties that experienced freeway opening during a particular decade had the highest rates of net migration during the next decade. Overall, construction of the IHS facilitated an increase in migration.

Through these findings, Briggs substantiated the work of Lichter and Fuguitt (1980) that modeled the functional linkages between population size, distance to the nearest metropolitan area, employment, and eventual migration (Figure 4.2). Link One indicates that freeway construction leads to increased employment in manufacturing, tourism, and non-local services. Link Two

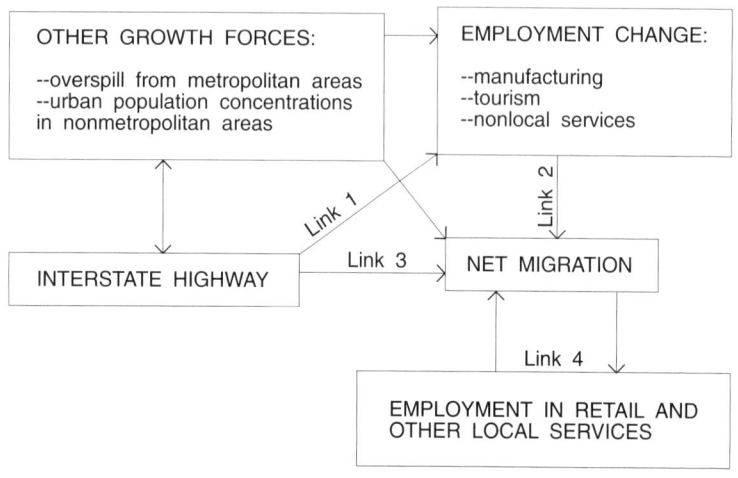

Figure 4.2. Model of IHS effects on net migration. Source: Lichter and Fuguitt 1980. Reprinted with permission of the University of North Carolina Press.

represents the migration of people toward these increasing employment opportunities. Link Three represents increasing freeway commuting from suburban and exurban communities. Link Four denotes a second wave of migration occurring as service sector employment opportunities increase as a result of earlier migration. This second wave signifies the circular relationship between increasing migration and increasing employment.

Eyerly, *et al.* (1987) concurred that the IHS has increased employment opportunities in nearby nonmetropolitan places. They examined a number of traditional economic indices as well as several of their own creation. First, they examined the network's impact on population, income, housing, and employment. Counties with interstate highway routes were developing at an increased rate compared to their counterparts. All of the indices examined substantiate the strong growth, especially industrial growth, occurring around nonurban interchanges when compared to county and state growth levels (Eyerly, *et al.* 1987). In addition to testing for growth among these more traditional measures, they also tested several indices of the assessed market value of real estate. Again, they found a positive correlation between the network's presence and economic growth (Table 4.3).

TABLE 4.3
Percent Change in Selected Economic Development Indices
for Pennsylvania, 1970-1984

Development Type	State	Counties on the IHS	Interchange Communities
Residential	35%	44%	64%
Commercial	21%	44%	95%
Industrial	-13%	9%	90%
All Types	27%	39%	65%

Source: Eyerly, et al. 1987.

Rather than measure the positive spinoffs of proximity to a rural interstate highway, Baltensperger (1991) focused on the negative aspects of isolation. The value of a rural county's location in an interstate highway corridor was documented in a study of farm population in the Great Plains. The study examined rural counties currently missing the so-called "rural renaissance" noted in some developing areas. These counties were generally in more remote areas, considered to be headed "downhill" and suffering a lack of the *"prerequisites of growth"* (Baltensperger 1991, 441). These prerequisites include the presence of a large central place and a location near the IHS.

The Network's Role in Economic Restructuring

Bell and Feitelson (1991, 518), an economist and a geographer, clearly and succinctly state the relationship between an economy and transportation infrastructure:

> *Physical infrastructure facilities are not ends in themselves, but rather they interact with private capital to produce a service. In the transportation area, public capital (roads, bridges, runways) interact with private capital (cars, trucks,*

airplanes) to produce a service (mobility). Thus, public capital affects economic activity by complementing, rather than substituting for, private capital. In this context, transportation services can be considered an intermediate good in the private production and consumption process, and demand for transportation services is a derived demand that depends on the demand for final products.

Twentieth century manufacturing development involved the relocation of industry to suburban and peripheral areas (Cohen and Berry 1975). As confirmed by Fisher and Mitchelson (1981), this is especially relevant in nonmetropolitan areas linked to the global economy via the IHS.

Erickson (1981) points out that nonmetropolitan employment in manufacturing is growing at a rate far greater than that exhibited in metropolitan areas. Four changes in industrial organization that contribute to this process are: [1] existing operations expand into nonmetropolitan areas; [2] new firms develop from within nonmetropolitan areas; [3] plants relocate from metropolitan to nonmetropolitan areas; and [4] branch plants. Each of these factors is vitally linked to the IHS, especially the fourth. Branch plants arise when larger firms attempt to separate spatially the corporate decision making function from the manufacturing process. While a corporation's headquarters might remain in a large metropolitan area, a network of smaller manufacturing centers may be spread around the nation and/or world where labor or other cost savings take place. Connections between interstate highways and this industrial "filtering down" process into nonmetropolitan areas are further documented by Cromley and Leinbach (1981).

In a subsequent comparative look at the attitudes of manufacturers, Zemotel, *et al.* (1987) compare the transportation needs of traditional versus "advanced technology" firms. These firms are identified through SIC codes and feature a higher proportion of technical workers and research and development expenditures. Their transportation needs, identified through a survey, are different from those of traditional firms. Advanced technology firms:

[1] prefer greater regional accessibility, especially via interstate highways, state highways, and airports;
[2] rely more on express parcel delivery;
[3] place a higher priority on air transport for both personnel and products;

[4] more often cite highway problems as concerns; and
[5] frequently cite traffic problems as sources of worry.

In addition, high-tech firms reportedly preferred industrial park locations especially near interstate highways.

Regional Development in the I-75 Corridor

An interesting case study of one interstate highway and its role in economic development is I-75. The I-75 corridor extends from the International Bridge between Sault Ste. Marie, Ontario, and Sault Ste. Marie, Michigan, and a point in south Florida near Miami. This 1,600 mile link of the IHS transects Michigan, Ohio, Kentucky, Tennessee, Georgia, and Florida. Transcontinental access via I-75 is enhanced by its intersections with major east-west interstate routes at Toledo (I-80 and I-90), Dayton (I-70), Knoxville (I-40), and in north-central Florida west of Jacksonville (I-10).

Recent estimates of U.S. capital spending on manufacturing across the country indicate that as much as 15 percent occurs within the I-75 corridor (Finn 1987, 80). Finn also estimates that roughly 25 percent of Japanese investments in the U.S. occur within the corridor. With a mixture of automobile assembly plants and suppliers in combination with defense contractors, the I-75 corridor might be viewed as a transnational contradiction. Scattered among midwestern cornfields, Kentucky horse farms, and Georgia pine forests are the I-75 signs spawned by General Motors, Mazda, Honda, General Electric, Honeywell, GTE, Citicorp, Toyota, Ford, Martin Marietta, Komatsu, Lockheed, Northrup, and McDonnell Douglas. Among the reasons cited by Finn (1987) to explain the industrial clustering along the corridor are:

[1] population redistribution to the south and west;
[2] industrial decentralization;
[3] branch plant development;
[4] an increased reliance on suppliers;
[5] just-in-time manufacturing;
[6] heavy subsidies from individual states;
[7] an abundant supply of blue-collar,
    non-unionized labor;
[8] high land availability at low cost;
[9] central national and continental location;

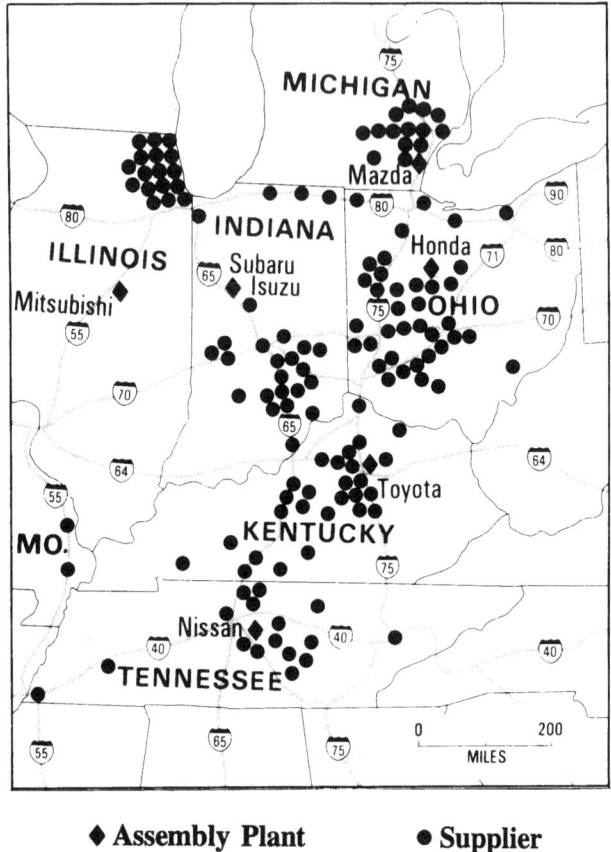

Figure 4.3. Japanese motor vehicle production in the I-65 and I-75 corridors. Source: Rubenstein 1990. Copyright The American Geographical Society. Reprinted with permission.

[10] reduced transportation costs; and,
[11] reduced costs of doing business and costs of living in general.

In a detailed examination of the U.S. automobile industry, Rubenstein (1990) found that the I-75/I-65 corridor remains the focus of U.S. and Japanese manufacturing investment (Figure 4.3). Within the realm of automobile manufacturers and suppliers, I-75 and I-65 are known as the *"kanban"* highways. *Kanban* is the

Japanese word for *"just-in-time"* delivery (Rubenstein 1990, 9). In fact, of the 20 motor vehicle assembly plants started in the U.S. between 1980 and 1991, 16 were built within 100 kilometers of the corridor (Rubenstein 1992). Rubenstein cites the driving force behind the production pattern's shift to this region and away from the Great Lakes as an ongoing effort to minimize transportation costs. And, according to the geographer, *"fragmentation of the U.S. motor vehicle market"* occurs as U.S. and foreign controlled manufacturers offer an increasingly wide variety of products disseminated through new assembly plants and supplier networks (Rubenstein, 1992, 433). This fragmentation in conjunction with large scale population redistribution, regional labor problems/ costs, and the availability of the IHS has redefined the pattern and process of U.S. automobile assembly and distribution. The cumulative result of these processes is a well-defined, regionally specific industry that is deeply rooted alongside I-75 and I-65.

The Arkansas Example

Arkansas is transected by a major transnational IHS route, I-40, and segments of I-30 and I-55. The state is comprised of 75 counties, 20 counties with IHS routes and 55 without. The *"land of opportunity"* is intriguing from a regional development perspective because of the noted changes taking place there during the last two decades and the recent national attention focused on the state.

A county level examination of Arkansas reveals two divergent patterns of regional development. By examining the changes in two sets of counties, those with and without IHS routes, the advantages of a route's presence become clear. Table 4.4 contains a list of attribute variables collected for the two sets of counties. IHS counties feature distinct population, economic, and employment dynamics. Supporting the work done in Pennsylvania by Eyerly, *et al.* (1987), the Arkansas employment and payroll patterns present an increasingly clear image of IHS advantages. As Baltensperger (1991, 441) suggests, counties without interstate routes could be "headed downhill."

Commuting Processes and Patterns

Employment impacts of the location of economic activity related to the IHS and discussed in the previous section have important implications for commuting. A laborshed envisioned as a com-

TABLE 4.4
Post-IHS Demographic and Economic Attributes
of Arkansas Counties

Mean County Change in:	Counties on the IHS	Counties not on the IHS
Total Population, 1960 - 1990	13,805	5,243
Total Personal Income (in thousands), 1970 - 1989	$592,459	$253,738
Public School Enrollment, 1960 - 1990	299	-293
Manufacturing Employment, 1960 - 1990	2,430	1,430
Manufacturing Payroll, 1960 - 1990	$88,668	$44,858
Total Payroll (in thousands), 1960 - 1990	$416,689	$133,372
Total Bank Deposits (in thousands), 1960 - 1990	$334,979	$177,955
Real and Personal Property Assessments (in thousands), 1960-1990	$279,977	$139,707
Retail Sales (in thousands), 1972 - 1987	$194,846	$69,708
Wholesale Sales (in thousands), 1972 - 1987	$299,223	$60,209
Service Establishment Sales Receipts (in thousands), 1972 - 1987	$94,513	$23,687

Source: Wayman and Hassell 1992. Calculations by Rob Breymaier.

muting zone is one type of hinterland surrounding a city and is defined by the transportation networks that employees use on their journeys to work. The journey to work is recognized as the most important of all trip types (Barber 1986). Commuting is one example of a spatial process that leads to the formation of a unique spatial pattern, the commutershed.

According to the American Association of State Highway and Transportation Officials, the United States is experiencing a "commuting explosion" (American Association of State Highway and Transportation Officials 1987, 1). In an analysis of commuting by Eno Transportation Foundation, Inc. (1987), a host of factors influencing post-War commuting patterns were considered, some of these noted in earlier chapters. The population of the United states increased by 50 percent between 1950 and 1980 while the number of workers grew by 65 percent during this same period. The post-War baby boom has had an impact on overcrowding of highways and the number and percentage of women in the workforce steadily increased (from 28 percent in 1950 to 42 percent in 1980). That the availability of private automobiles for personal use has nearly doubled since the end of World War II was also noted. In all but six states (Arkansas, Louisiana, New York, Nevada, Utah, and West Virginia) and the District of Columbia, the number of automobiles per capita exceeds 0.5 (see Figure 2.6). In addition, the number of households without vehicles has sharply declined during this same period. And for the most part, households without vehicles are small households without workers, predominantly located in the center cities of large metropolitan areas (Eno 1987). Rural to metropolitan migration has created more potential commuters (more than 75 percent of the U.S. population now resides in metropolitan areas). Urban to suburban migration has increased the length of the American commute and the dependence on the private automobile. Since 1950, more than 86 percent of national population growth has taken place in the suburbs -- roughly 44 percent of the U.S. population is now suburban (Eno Transportation Foundation, Inc. 1987).

To many Americans suburban life is equated with commuting, often in a private automobile carrying only one passenger -- the driver. The average private commuter vehicle now transports 1.15 persons per vehicle and that figure is in a downward trend (Eno Transportation Foundation, Inc. 1987). Shifts in the location of employment have accompanied the urban to suburban migration of residences. The number of suburban jobs grew to nearly 50 percent of all those located in metropolitan places between 1960 and 1980. During this same period, roughly 67 percent of all metropolitan job increases occurred in suburbia.

At the national level (Figure 4.4), the predominant commuter flow is suburb to suburb, twice the rate of that from suburb to central city (of the national increase in commuting between 1960 and 1980, roughly 83 percent occurred in the form of suburb to suburb travel). Commuting between metropolitan areas is increasing as is commuting from and to metropolitan areas. Both of these phenomena contribute to growing trip length and time. The report identifies the main theme of increasing suburban commuting growth as *"a trend toward greater balance between jobs and resident workers. As these ratios approach 1.0, the potential for commuting to remain within home jurisdictions increases, promising opportunity for commuting efficiencies"* (Eno Transportation Foundation, Inc. 1987, 38). Between 1960 and 1980, every alternative mode of commuter travel declined in usage relative to the private automobile. At the same time, the average length of the commute to work via the private automobile has grown to almost 22 minutes and roughly ten miles each way, resulting in an average speed of only 29 miles per hour. Interestingly, Eno (1987) reports that both travel distance and time for suburb to suburb commutes are roughly half that of their suburb to central city counterparts.

The IHS has also had an impact on commuting in nonmetropolitan areas. In an early study, Thiel (1962) presented a detailed spatial examination of commuting to one firm in New England. He cited a number of benefits often derived from freeway commuting by workers, including: more employment opportunities, time savings, money savings, convenience, and economical parking in freeway industrial parks. To take advantage of these benefits, some employees move further from work while prospective employees are attracted from greater distances. Their collective travel behavior results in the formation of a commutershed and maps its realm of influence. Thiel's map of two commuter fields around a single Connecticut manufacturing firm illustrates the nearby immediate spatial expansion of influence brought about by I-395 (the Connecticut Turnpike). Figure 4.5 compares Thiel's pre- and post-IHS commuter fields. Upon construction of the freeway, the axis of the study firm's commuter field increased from roughly 29 to 39 miles.

In a number of more recent regional analyses, Mitchelson and Fisher (1987a,b; 1988) further detail the expansion of metropolitan commuting fields. Following the work of Berry (1970) and others, their research illustrates the potential significance of capital transfer into nonmetropolitan hinterlands from larger urban centers via

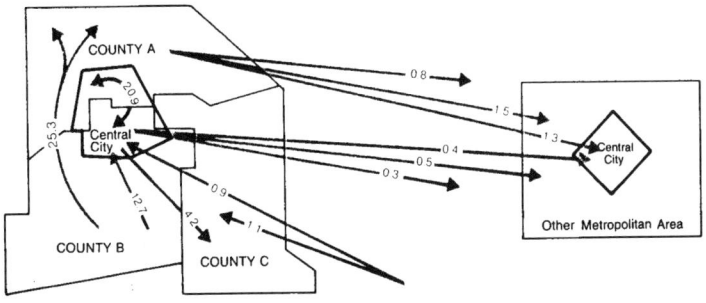

Figure 4.4. A model of U.S. commuting. Source: Eno Transportation Foundation, Inc. 1987. Reprinted with permission of Eno Transportation Foundation, Inc.

commuting. By examining commuting fields from 1960 to 1980, they substantiate the notion that intermetropolitan, peripheral regions benefit from the extended commuting fields of larger urban areas. Mitchelson and Fisher find that commuting is an "export industry" that attracts both people and capital to nonmetropolitan areas. Modern commuting fields exist up to 60 miles from some urban areas impacting the economies of nonmetropolitan communities. These commuting fields are expanding into more and larger nonmetropolitan areas, creating a complicated system of commuter fields around urban centers of all sizes. Income and population growth can flow from both metropolitan and smaller nonmetropolitan centers into nonmetropolitan areas, resulting in the reduced importance of agglomeration economies in modern America. "*Results clearly point to the permissive role that implantation of limited access highways [the IHS] play in development via commuting*" (Mitchelson and Fisher 1988, viii). They further support the growing positive evidence of economic influence of highways on nonmetropolitan communities, especially those located on freeways. There commuters join local residents and passers-through in strengthening rural markets for goods and services. However, the distribution of this process is not spatially uniform but focused on key nonmetropolitan places.

Conclusion

Analysts disagree over the costs, benefits, and economic spinoffs of freeway investments. Future population, social, and economic

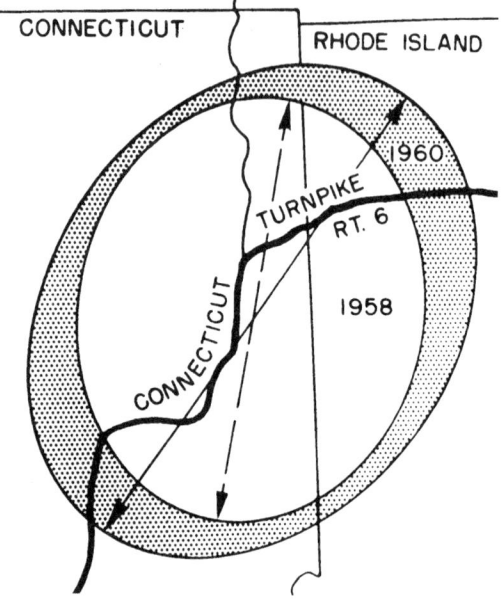

Figure 4.5. Pre- and post-highway commuter fields. Source: Thiel 1962.

changes are likely to further fuel those debates. Even among debaters, however, there is agreement on one critical point. Highway development facilitates regional development. The degree to which the IHS has gone beyond simple facilitation to redefine large areas of the U.S. remains uncertain (Table 4.5).

Among the many factors that increase the difficulty of understanding the relationship between transportation investment and return are scale and timing. When viewed and evaluated from a national perspective, IHS expenditures are measured against a different set of criteria than at the local or regional scales. Economists carrying out national studies of costs versus benefits and the impact of freeway expenditures on other modes cannot always adequately factor in community level impacts. At the same time, regional economic development experts citing long lists of local success stories of communities thriving from their proximity to interstate routes cannot effectively gauge the national costs of local growth.

TABLE 4.5
Top Ten States by IHS Designated Mileage

Rank	State	IHS Mileage
1.	Texas	3,162.58
2.	California	2,311.51
3.	Illinois	1,715,10
4.	Pennsylvania	1,563.85
5.	Ohio	1,530.39
6.	Florida	1,460.19
7.	New York	1,392.55
8.	Georgia	1,199.90
9.	Montana	1,187.54
10.	Michigan	1,174.00

Source: U.S. Department of Transportation, Federal Highway Administration 1992a.

Perhaps more importantly, neither critics nor advocates of the IHS have generally taken a longitudinal approach to analysis of IHS costs and benefits. How long does a highway system have to function before its contributions to the local, regional, or national economies are fairly assessed? Many of the studies finding little evidence of regional economic development via interstate highways were completed before the system was a truly functional network. They were carried out during Wilson's Phase One and probably minimize the network's impact. Later studies, citing the prematurity of the 1950s and 1960s research, reach a different conclusion. Perhaps we have reached a point in IHS history when a more comprehensive analysis of its impacts across a variety of scales can be conducted.

5

IHS INTERCHANGES

> *Interstate interchanges are becoming oases around which cities are being built. Urban areas are sprouting where corn or wheat or cotton was king only a few short years ago. (American Trucking Associations 1967, 73)*
>
> *A great motorway has no business cutting a wide swath right through a town or city and destroying the values there; its place is in the country where there is ample room for it... (Norman Bel Geddes quoted in Patton 1986, 100)*
>
> *The nation that lives on wheels still has the dubious honor of having created, along 3,000,000 miles of highway, the supreme Honky-Tonk of all time. (Life Magazine quoted in Patton 1986, 66-67)*

While the Interstate Highway System's impact is clear at the national and regional scales, it is also manifest in intensive local land use change around the system's interchanges. This is especially true in areas less developed prior to highway construction where existing land uses were perhaps less resistant to change. In many densely settled urban areas, there remains very little developable space around IHS interchanges. In New York City, for example, several interchanges were squeezed into the existing land use pattern (Figure 5.1). In most rural areas, the existing land uses are less intensive -- primarily agricultural -- or land parcels are vacant. Consequently, development is easier, and a logical measure of the system's micro-level impact in rural and nonmetropolitan areas.

Until recently, land use change in nonmetropolitan areas was neglected in the literature, particularly in comparison to studies of

90 IHS Interchanges

Figure 5.1. IHS interchanges near the George Washington Bridge in New York City.
Source: Lockwood, Kessler & Bartlett, Inc. 1992.

urban and suburban areas. *"The most recent research on land use change has focused on: conversion in the urban-fringe, the decentralization of retail and industrial activity, commercial land use succession, redevelopment in the central city, and aggregate land use change"* Wilder (1985, 332). To address this gap, this chapter focuses on the relatively recent research on interstate highway interchanges as instigators of nonmetropolitan land use change.

Interchange Communities

Logically, the quantity and quality of development generated by IHS interchanges might be expected to vary from place to place depending on a variety of site and situation characteristics. Twark (1967) discovered that most new economic development at nonmetropolitan interstate highway interchanges occurred within 0.5 miles of the intersection. Consequently, he defined the area within that distance of the crossroads as the "interchange community" and established its usefulness as an appropriate study area for interchange analysis. Another rationale for defining a circular study area is that linear strip analysis tends to de-emphasize development on roads parallel to the interstate highway. Square or rectangular study areas exaggerate the importance of changes in the "corners" which are a greater distance from the immediate influence of the interchange. A circular impact area with a radius of 0.5 miles encompasses 502 acres. The important locational attributes of high visibility and easy access available at IHS interchanges draw developers to these intersections. The Antelope Interchange on I-5 near Red Bluff, California, provides a clear example of new interchange community development and the attractiveness of at-ramp locations (Figure 5.2).

In documenting the new development at 105 interchanges located in nonmetropolitan regions of Pennsylvania, Twark (1967) observed that most of the new development around the interchanges was commercial, including mostly traffic-oriented establishments like service stations, restaurants, and motels. Residential development was the second most prevalent land use in new interchange community development, with new residential development evident in 40 percent of the interchange study areas. Little new industrial or institutional development occurred.

In order to explain patterns of interchange development, Twark tested two sets of independent variables, endogenous and exogenous, for their ability to predict interchange community

92 IHS Interchanges

Figure 5.2. The Antelope IHS interchange near Red Bluff, California. Source: American Aerial Surveys, Inc. 1990.

development. Endogenous variables measured the *"state or levels of economic development for the given community"* while exogenous variables were those that *"affect the level of economic development of the interchange community but which are not, in general, affected by the growth that takes place."* The most significant independent variable explaining new development was found to be average daily traffic (ADT), but on the cross route rather than the interstate highway itself. Other factors positively correlated with new interchange development included ADT on the interstate route, local population, topography, and distance from the nearest urban center.

While Twark's analysis failed to produce a well-defined model of the whole interchange community development process, it challenged previous research on IHS-generated land use change. First, Twark documented that new development was, in fact, occurring around certain nonmetropolitan IHS interchanges. Second, he challenged the conventional wisdom that interstate highway ADT is the sole criterion capable of explaining interchange community development.

Interchange Morphology

In a survey analysis of IHS interchange areas conducted twenty years later, Norris (1987) established the clusters of development found there as key components of the American cultural landscape. He documented the "morphological regularity" of the nonmetropolitan IHS interchange and placed it as a type alongside *"small town main streets, downtown cores, and suburban shopping plazas"* (Norris 1987, 23). Interchange communities were recognized as *"built forms and assemblages characterized by opportunism, obsolescence, and mutation"* (Norris 1987, 23).

The study analyzed 354 interchanges along I-75 between upper Michigan and southern Florida (Table 5.1). Interchange areas and their levels of development were ascertained through published directories of I-75 establishments. Of the 354 interchange communities that Norris extracted, 302 (85.3 percent) were categorized by him as "developed." More than 75 percent of the study interchanges in each of the six transected states were in this category: in Michigan, 77.8 percent; Ohio, 82.5 percent; Kentucky, 87.9 percent; Tennessee, 94.3 percent; Georgia, 86.0 percent; and Florida, 92.9 percent. Collectively, the 354 interchanges were surrounded by 2,618 commercial establishments. Gasoline stations, motels, and restaurants dominated the interchange communities, account-

TABLE 5.1
Levels of Development at I-75 Interchanges by State

State	I-75 Mileage	Number of Undeveloped Exits	Number of Developed Exits	Number of Establishments	Number of Establishments per Exit
Florida	211	3	39	274	7.0
Georgia	355	14	86	720	8.4
Tennessee	162	2	33	229	6.9
Kentucky	192	4	29	361	12.4
Ohio	210	11	52	467	9.0
Michigan	394	18	63	567	9.0

Source: Norris 1987.

ing for 75.4 percent of the commercial establishments. The bulk of these establishments were franchises or other outlets of national or international firms (Table 5.2). This development pattern represents an infusion of outside capital into these nonmetropolitan areas and potential competition for existing local firms.

Norris used the interchange community maps published in the directories to assess and generalize interchange morphology. Eight distinctive development patterns -- five "basic" and three "hybrid" -- could be ascertained. Norris' models of axis-oriented development (Figure 5.3) highlight the influence on IHS interchange development arising from the intersecting route.

Through his research on nonmetropolitan IHS interchange morphology, Norris went beyond Twark's work in documenting the significant levels of land use conversion occurring around certain interchanges. The data he used, however, were inherently biased to represent only commercial development, and he was forced to neglect noncommercial players in the interchange community dynamic. Even some regionally oriented commercial land uses, like the periodic flea market or swap meet, are missed, although the earlier work by Twark found very few of these in Pennsylvania. At this point, the role of the network in attracting noncommercial development to nonmetropolitan interchanges

TABLE 5.2
Commercial Establishments at I-75 Interchanges

Type of Establishment	Number	Percent
Gasoline Stations		
Major Oil Company	730	28.1
All Other Stations	231	8.9
Motels and Motor Hotels		
Major Chains	179	6.9
All Other Motels	187	7.2
Eating Establishments		
Fast Food Chains	203	7.8
Restaurant Chains	167	6.4
All Other Establishments	263	10.1
Other Services		
Retail Outlets, Plazas, Malls	357	13.7
All Other Roadside Services	281	10.8
Total	2,598	99.9

Source: Norris 1987.

remains unclear. The two decades that elapsed between these two studies was an important period in the devleopment of the system. During the early 1960s, the IHS was more a set of partially connected links than an operative network.

Toward a Model of Interchange Development

A 1986 study of variations in interchange development in Kentucky examined land use change around 65 nonmetropolitan IHS interchanges (Moon 1986a). In this research, each of the study interchanges was examined before and after IHS construction using aerial photographs of the interchange areas supplemented with field checking. Every interchange in Kentucky was studied in the analysis rather than including only those predetermined to be "developed," a practice used in earlier studies. Kentucky was an interesting study area based on its geographic diversity (physical and human), central location, east-west and north-south IHS

96 IHS Interchanges

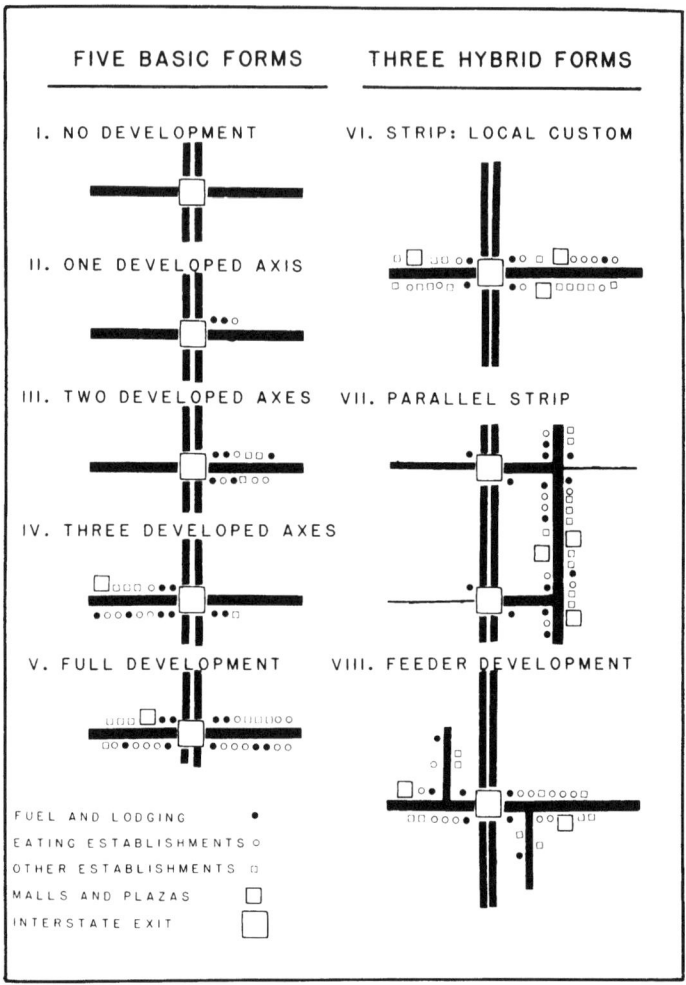

Figure 5.3. IHS interchange morphology. Source: Norris 1987.

routes, and interchanges of different geometric designs and ages. The study area around each interchange was circular with a radius of 0.5 miles, after Twark.

The level of new development around each interchange was identified by comparing existing development at the time of the study mapped from field research with development recorded in pre-construction photographs of the interchange areas. Differences substantiated the levels of interchange development as well as the diversity of developers (Table 5.3 and Moon 1987c). The average number of pre-construction structures per interchange community was just 28.1 compared to a post-construction count of 102.6 structures. In constrast to other studies of interchange development, the bulk of the new development observed in Kentucky occurred in the form of single-family residences, simple residence-related structures, and small commercial/small institutional buildings.

A development complexity score was assigned to each study interchange community based on the level of its development at the time of the study. As a measure of interchange community land use activity, the score incorporates the size and scope of structures

TABLE 5.3
Average Number of Structures Per Interchange Community in Kentucky, Pre-Construction and in 1985

Type of Structure	Pre-Construction	Post-Construction in 1985
Single-Family Residential	14.37	47.43
Multi-Family Residential	0.14	3.00
Simple Non-Residential	12.77	32.65
Small Commercial/ Small Institutional	0.14	2.83
Large Commercial/ Large Institutional/ Small Industrial	0.71	16.57
Large Industrial	0.00	0.14
Total	28.13	102.62

Source: Moon 1987b. Reprinted with permission.

TABLE 5.4
Interchange Community Complexity Score Contributions
by Structure Type Category

Category	Structure Type	Percent
I	Simple, Non-residential	10.08%
II	Single-Family Residential	29.29%
III	Multi-Family Residential	3.50%
IV	Small Commercial/Institutional	40.94%
V	Large Commercial/Institutional, Small Industrial	14.82%
VI	Large Industrial	1.37%

Source: Moon 1988. Reprinted from Land Use Policy, Vol. 5, H. Moon, "Modelling Land Use Change Around Interstate Highway Interchanges," pp. 394-407, Copyright 1988, with kind permission from Elsevier Science Ltd., The Boulevard, Langford Lane, Kidlington OX5 1GB, UK.

located at interchange sites and is a measure of the human and vehicular activity that the particular facilities generate. Because structures can vary widely by size and uses, a weighting algorithm allowed a composite complexity score to be assigned to each study area depending on the number of structures and their membership in one of six structure categories. The larger and more complex a structure was, the greater its weight.

Among the 65 interchange communities studied, the calculated complexity scores ranged from 13 for an interchange community in Marshall Couny to 1,616 for one near Paducah (Moon 1988, 399). The interchange study area in Marshall County featured eight structures: five single-family residences and three simple, non-residential structures. The interchange near Paducah was surrounded by 247 structures of a variety of sizes, forms, and functions. The importance of each structure type across all interchanges is shown in Table 5.4. Given the information on variation in the numbers and types of structures developed in the interchange community, the next step was to model the interchange community development process by focusing on independent variables associated with fostering or retarding new development.

The Interchange Development Model

A key to modeling interchange-specific development is the documented variability in interchange community complexity scores. In all, 22 independent variables were tested for their association with variations in development complexity (Moon 1988). These included highway variables such as interstate ADT and age of the interchange, site variables like topography, size of developable area at the interchange site, and number of pre-construction land owners, and situation variables like distance to the nearest Metropolitan Statistical Area and distance to the nearest neighboring interchange. An attribute of special relevance in Kentucky and other states is the status of the local community as a "wet" or "dry" community with regard to the sale of alcoholic beverages.

A stepwise regression procedure combining forward and backward entry was used to construct models of interchange development complexity. The following independent variables were the set that best explained variations in interchange community development complexity:

[1] preconstruction development complexity (+);
[2] restrictive interchange design type (-);
[3] location in an area designated to sell alcoholic beverages (+); and,
[4] location in a high latent demand area (+).

Collectively, the four independent variables explained slightly more than 53 percent of the variation in the dependent variable, current interchange development complexity.

Analysis of the residuals of this modeling effort indicated some interesting patterns. The study areas around the seven interchanges located within the corporate limits of larger cities (populations of 25,000 or more) were generally more developed than the remaining areas and the model underestimated the level of their development complexity. Removal of these interchange areas increased the explanatory power of the model but converted it from a nonmetropolitan interchange development model to a nonurban one. The significant independent variables in the model for interchanges outside of the larger cities were:

[1] preconstruction development complexity (+);
[2] location in a high latent demand area (+);

[3] distance to the nearest neighboring interchange (+);
[4] IHS ADT (+);
[5] location in an area designated to sell alcoholic beverages (+);
[6] distance to the nearest city of any size (-); and,
[7] topography (-).

The land use conversion process documented in this study was a complex one. Development patterns were different for interchanges within larger cities. For smaller cities, development complexity was a function of interstate attributes and interchange site and situation characteristics.

Interchanges as Regional Growth Poles

The regional importance of IHS routes was described in the previous chapter. In an investigation of recent changes in Appalachia, the IHS was identified as a key factor (Moon 1986b). A set of interchanges lying along I-64 and I-75 in Appalachia were compared in the extent of their development to a set of interchanges on I-24, I-64, I-65, I-71, and I-75. In comparing and contrasting local development around the 19 IHS interchanges in Appalachia with that around 46 IHS interchanges located outside Appalachia, Moon found a number of differences. Table 5.5 illustrates some of the key differences found in the two sets of interchange communities studied.

Differences in spacing, relative location to urban areas, and the actual level of surrounding development were among the most outstanding. On average, interchanges in Appalachia are farther apart than interchanges outside of the region. Interchanges in Appalachia are also more removed from both small and large cities. The mean distance between an Appalachian interchange and a city of any size is 2.1 miles compared to a mean distance of 1.5 miles between non-Appalachian interchanges and similar places. In terms of their relative location to larger cities (those with populations exceeding 25,000), Appalachian interchanges are 23.8 miles farther from large cities than their counterparts. The average distance between an Appalachian interchange and a city of 25,000 of more is 46.5 miles compared to just 22.7 miles between non-Appalachian interchanges and similar places.

The locational advantages held by Appalachian interchanges as key nodes in an underdeveloped region are particularly appar-

TABLE 5.5
Differences between IHS Interchanges Located Within and
Located Outside of Appalachia

Interchange Attribute	Interchanges Within Appalachia	Interchanges Outside Appalachia
Pre-construction Complexity	40.9	53.7
Total Complexity	416.4	281.5
Residential Complexity	120.2	92.9
Non-residential Complexity	261.9	148.7
Interstate Highway ADT	20,685	17,821
Interchange Age (in years)	18.6	16.3
Nearest Neighboring Interchange (in miles)	4.4	3.8
Farthest Neighboring Interchange (in miles)	10.1	7.3
Miles of Primary Highway in Interchange Study Area (per square mile)	0.05	0.09
Miles of Primary Highway in Interchange Study Area (per 1,000 people)	0.65	1.10

Source: Moon 1986b. Reprinted with permission.

ent in the commercial, institutional, and small industrial sectors. Although more single- and multi-family residences exist in Appalachian interchange areas than in those outside the region, there is a wider disparity between the two sets of interchanges in terms of nonresidential development. In the case of small scale nonresidential development, Appalachian interchanges had an average of 24.1 structures compared to a mean of 13.2 for interchanges outside the region. Appalachian interchanges feature an average of 4.1

large commercial, large institutional, or small industrial structures while their counterparts have just 2.4. In total, Appalachian interchanges are surrounded by an average of 119.6 structures while non-Appalachian interchanges are encircled by an average of 101.6, a difference of almost 18 percent. In fact, the only structure type not more commonly found around Appalachian interchanges is large industrial. An examination of these 65 study areas found significantly less development in the 19 Appalachian cases before IHS construction. The operation of IHS interchanges as regional growth poles is most evident at the network's intersection with the Appalachian Regional Commission's Appalachian Development Highway System (Moon, 1986b).

An Interchange Community in the West: Little America

When early pioneers trekked west along Nebraska's Platte River past Chimney Rock *en route* to California, Oregon, or Utah, they invariably frequented the Robidoux Trading Post. Joseph Robidoux's combination blacksmith shop, saloon, and outpost near Scott's Bluff served as the *"jumping off place"* for the next and most difficult leg of the westward trip, crossing the Rocky Mountains. This key "interchange" was located in a gap that became known as Robidoux Pass and was strategically sited near a rare source of fresh water and firewood. Between 1849 and the early 1850s, most westward-bound wagon trains passed through Robidoux Pass and patronized the trading post there (Mattes 1983). Later, the preferred route west would shift five miles to the north through Mitchell Pass. There the Pony Express, Overland Stage, and the first transcontinental telegraph would foreshadow the "communication bundle" that remains today. A critical period in U.S. history, a finagling entrepreneur, and an optimal location alongside a transnational route coincided to create a node, an interchange, a new place.

Today, a number of parallels to Robidoux's Trading Post exist alongside the major interstate highway routes. Most do not go by the label "trading post" but they do function along the same lines, meeting the basic needs of travelers. While perhaps not as unscrupulous as Joseph Robidoux, the modern interchange village entrepreneur also uses location to his or her advantage. A different model of the future interchange trading post is evolving at a number of key western sites. These new models of large scale

Figure 5.4. The IHS interchange village at Little America, Wyoming. Source: Worden 1992b.

commercial success are all based on the same prototype -- Little America, Wyoming. Interestingly, this confluence of spatial activity lies just west of Scott's Bluff.

Little America was started in desolate western Wyoming in 1936 by Steven S. Covey, a local businessman infatuated with Admiral Byrd's Antarctica exploits (Worden 1992a). Equating his surroundings with the frozen continent, Covey followed the lead of the Admiral and named his outpost "Little America." This place would become renowned as a respite, not only from the cold but from the hot, dry Wyoming summers. The service station/restaurant/garage would remain virtually unchanged even after Covey was forced to move his business to its current site following a devastating fire. The move some six miles to the southeast was a fortuitous one, because I-80 would later be built on Covey's doorstep.

Today, Little America reigns as the preeminent interchange service center offering a full menu of services to passersby. The combination service station/truckstop/restaurant/motel controls over 600 acres of land on the northern side of Interchange 68 between Cheyenne and Salt Lake City (Figure 5.4). There it meets the needs of nearby soda ash and petroleum industry firms and employees in addition to those of its residents and scores of I-80 truckers and travelers. A number of unique circumstances characterize Little America because of its relatively isolated location, including a summer employment peak near 260 (140 full-time employees and their families live in company-owned homes on the site and another 40 employees are provided on-site apartments). Employees are bused up to 30 miles from three adjacent communities while the Little America water supply is pumped six miles to the site. Solid waste is regularly trucked some 30 miles for disposal and liquid waste is treated in an on-site waste lagoon. Little America also has its own post office and designated zip code and a full-time fire crew. In this desolate environment where employees routinely rescue elk and cattle from the complex's swimming pool, 1.5 million gallons of diesel fuel have been pumped in a single month (Worden 1992a).

As a complex spatial system that performs much like an oasis, Little America plays a number of vital functions. In July and August, it operates as an oasis for overheated RVs and station wagons. In deep winter, it acts more like a retreat when I-80 is closed during snowstorms. Between Easter and Thanksgiving, it is a tourism and honeymoon hotspot situated between Flaming

Gorge and Yellowstone National Park. Plans are in the works for two additional phases to include a larger convenience/grocery store and an RV park. In western Wyoming, Little America has no rival and the capital generated there has spawned a number of other investments. Little America-generated capital has financed three other similar outposts (in Cheyenne, Flagstaff, and Salt Lake City), the Little America Hotel in Salt Lake City, the Westgate Hotel in San Diego, and the ski resort at Sun Valley, Idaho (Worden 1992a).

Conclusion

IHS interchanges are key nodes connecting the network to the nation. But, they fulfill a role equally important to providing local entrance and egress. Interchanges provide new locations, new sites, new opportunities for economic and social activity (Table 5.6). They connect the IHS to the various regions of the U.S. and allow access to previously untapped places. The positive and negative attributes of the IHS are inevitably focused on the areas around its interchanges.

TABLE 5.6
Top Ten Interstate Highway Interchange Villages

Village	Interstate
1. Little America, Wyoming	I-80
2. Breezewood, Pennsylvania	I-70
3. Baseball City, Florida	I-4
4. Williamsburg, Iowa	I-80
5. Tucumcari, New Mexico	I-40
6. South of the Border, South Carolina	I-95
7. Wall, South Dakota	I-90
8. London, Kentucky	I-75
9. Christiansburg, Virginia	I-81
10. Stony Ridge, Ohio	I-280

Many questions are raised by the research on interchanges cited in this chapter. Is interchange community development imported from outside or is it merely borrowed from older nearby settlements? How will the political issues that eventually surface around certain interchange communities be addressed as they grow from simple clusters of development to central places? Roughly 10 percent of interchange communities in Kentucky are now recognized as interchange villages, central places functioning as commercial, industrial, administrative, institutional, and social nodes (Moon 1989). Is the development process that creates interchange villages parallel to observed urban development processes under which small towns grew up at key crossroads or are different processes at work in interchange communities? How will local, state, or federal agencies attempt to solve the law enforcement, sanitation, and traffic problems occurring around some of these hyperactive centers? These and other interesting questions demand additional scholarly research so that more can be understood about the local land use implications of the network.

6

Epilogue

A number of lessons can be drawn from analysis of the last 100 years of highway funding (Parker 1991). First, highway funding is rarely equitable between states and contributes to a *"winners and losers"* regional redistribution of wealth. Second, Congressional interest in major programmatic restructuring wanes immediately after passage of a major authorization bill. Third, Congress has a strong incentive for continuing its infatuation with the highway program and the power that accompanies it because the popular public works programs provide major "pork barrel" projects for Congressional constituencies. Fourth, the needs of many beneficiaries of the current highway and transit programs must be included in development of any new programs. Finally, broad, sweeping changes in transportation programs are generally less likely to occur than incremental changes.

Parker's "lessons" explain why, how, and when *The Intermodal Surface Transportation Efficiency Act of 1991* (ISTEA) was enacted (U.S. Congress, 1991). When President Bush signed ISTEA into law near the end of 1991, the federal authorization and allocation process was established for at least the next six years. The Act's $155 billion dollar price tag covers FY 1992-1997 and its statement of intent is clear:

> ...to develop a National System that is economically efficient, environmentally sound, provides the foundation for the Nation to compete in the global economy and will move people and goods in an energy efficient manner (U.S. Department of Transportation 1992, 5).

In terms of the IHS, ISTEA minutely changes the authorization and allocation patterns established over the decades of U.S. highway transportation policy. Its main impacts on the IHS center around the formation of an elite set of roads designated the National Highway System (NHS). These highways of "national significance" are set aside to receive privileged fiscal and planning priority. NHS's 155,000 miles include all of the IHS, a high proportion of urban and rural U.S. routes, the defense strategic network, and the most vital highway connectors. The NHS garners 13.5 percent ($21 billion) of the ISTEA authorization for just slightly less than 4 percent of the nation's 3.9 million miles of roads (U.S. Department of Transportation 1992, 8).

In addition to the funds set aside through the NHS, the IHS has access to other monies. These include $7.2 billion for IHS completion, $0.96 billion for substitution projects, and $17.0 billion to resurface, rehabilitate, restore, and reconstruct (the four "R's" of transportation improvement) the IHS (U.S. Department of Transportation 1992, 8). It should be noted that under the $17.0 billion allotment, reconstruction can only take place if it does not increase capacity other than through the addition of high occupancy vehicle (HOV) lanes. ISTEA includes a donor equity authorization to address the spatially uneven pattern of Highway Trust Fund allocations discussed in Chapter Four. ISTEA is an extension of traditional U.S. highway-dominated transportation policy that really does little to change the process and pattern of U.S. government spending on transportation. As illustrated in Figure 6.1, the highway spending slices of the ISTEA pie still dominate all others. When combined, spending on the NHS and IHS and on other highways and bridges commands more than half of the authorization.

In spite of the evidence that the costs of transportation innovation are high compared to the economic, social, and environmental benefits these innovations generate which are often lost in the ever-changing and complex world, politicians continue to endorse more and more costly transportation expenditures. Although America's interstate highway system is deteriorating, and traffic congestion in its urban centers is worsening, policy discussions about the road system are in gridlock because the only consensus that seems to have emerged is that public spending must be increased (Small *et al.* 1989). Increased budget authorizations and allocations are only made easier by the absence of a true national transportation policy. Dearing and Owen (1949) stated that our national leaders could not even agree on a goal for our national transportation systems. Federal transportation programs have not always considered

Epilogue 109

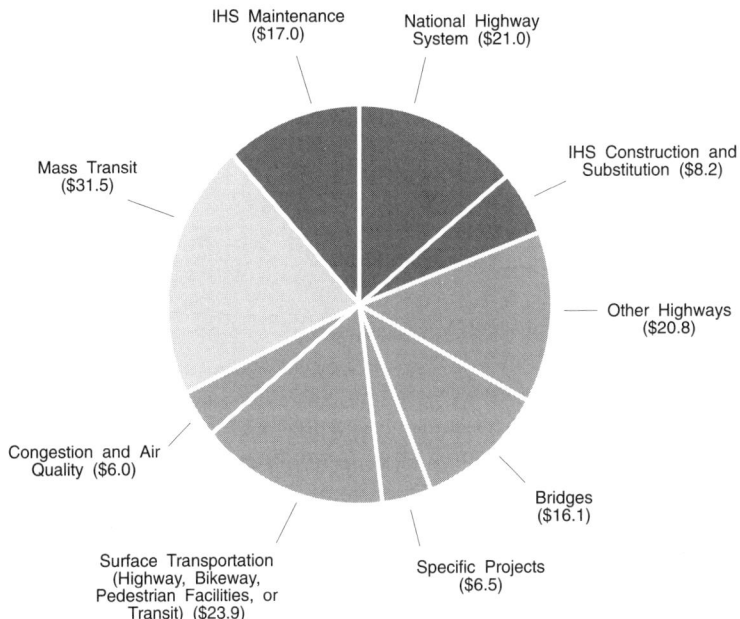

Figure 6.1. ISTEA authorization by type of expenditure in millions of dollars. Source: Plous 1993.

achieving the best possible transportation system for the movement of people and goods as the prime policy objective.

While the U.S. lacks a national transportation policy, we continue to travel. In Chapter Two it was reported that U.S. citizens purchased more than 114 billion gallons of gasoline, registered more than 188 million vehicles, and drove more than 2 trillion miles in 1990 (U.S. Department of Transportation, Federal Highway Administration 1991b, 6,17). Lowry (1988) expects these trends to continue well into the twenty-first century (Table 6.1). His figures, in conjunction with Reno's 1988 data on automobile ownership and transportation expenditures, forecast a gloomy future for urban freeways. These projections at least partially explain why so many metropolitan areas are seriously considering the construction of more and larger highways (Walters 1992). With the demands currently placed on urban government coffers, travelers can expect to see more localized and specialized highway funding mechanisms. Local and state governments and possibly

TABLE 6.1
Projections of Local Travel by Private Vehicle, 1980-2020

	1980	1990	2000	2010	2020
Licensed Drivers	141	162	176	191	199
Number of Vehicles (in millions)	135	159	176	191	199
Vehicles per Driver	0.96	0.98	1.00	1.00	1.00
Number of Vehicle Miles (in billions)	1,434	1,676	1,814	1,933	1,975
Annual Miles per Driver	10,191	10,318	10,300	10,115	9,906

Source: Lowry 1988.

private investors may be called on as the U.S. experiences, ironically, a return to toll roads in one form or another as a solution to traffic congestion. Regardless of which route our transportation future takes, along the IHS or other networks, there is nothing in our history to suggest that the trip will be guided by a coherent national transportation policy.

Bibliography

American Aerial Surveys, Inc. 1990. *Black and white photo of Antelope Interchange.* ASC #53-9141, 2-Teh-5 print dated 7/5/90. Sacramento, CA: American Aerial Surveys, Inc.

American Association of State Highway Officials (AASHO). 1956. *A policy on design standards, interstate system.* Washington: American Association of State Highway Officials.

American Association of State Highway and Transportation Officials Subcommittee on Highway Transport (AASHTO). 1987. *AASHTO news.* Washington: American Association of State Highway and Transportation Officials.

American Trucking Associations, Inc. (ATA). 1967. Highways - the years beyond 1972. In *Modern transportation: Selected readings,* ed. M. T. Farris and P. T. McElhiney, p. 62-73. Boston: Houghton Mifflin Company.

Baerwald, Thomas J. 1978. The emergence of a new "downtown." *The Geographical Review* 68:308-18.

_____. 1982. Land use change in suburban clusters and corridors. *Transportation Research Record* 861:7-12.

Baltensperger, Bradley H. 1991. A county that has gone downhill. *Geographical Review* 81:433-42.

Banks, Ralph K. 1984. An historical overview: National defense and the national system of interstate and defense highways. *Public Works* 6/84: 74-79.

Barber, Gerald. 1986. Aggregate characteristics of urban travel. In *The geography of urban transportation*, ed. S. Hanson, p. 73-90. New York: Guilford Press.

Bell, Michael E. and Feitelson, Eran 1991. U.S. economic restructuring and demand for transportation services. *Transportation Quarterly* 45:517:38.

Berry, Brian J. 1960. The impact of expanding metropolitan communities upon the central place hierarchy. *Annals of the Association of American Geographers* 50:112-16.

_____. 1970. Commuting patterns: Labor market participation and regional potential. *Growth and Change* Vol.1, No.4:3-10.

Black leaders, congressman fight road project in Roanoke. 1992. *Virginia News and Daily Advance*, July 31, p. B-1, B-3.

Bloch, A. J. and Crowell, W. H. 1984. Block grant transportation financing: The interstate trade-in experience. *Transportation Research Record* 967:6-10.

Briggs, Ronald. 1980. *The impact of the interstate highway system on nonmetropolitan growth.* Washington: U.S. Department of Transportation, Research and Special Programs Administration.

California Transportation (CALTRANS) Photo Department. Undated. *File photo of the Embarcadero freeway.* Photographed by Steve Hellon. Oakland: CALTRANS.

Cervero, Robert. 1986. *Suburban gridlock.* New Brunswick, NJ: Center for Urban Policy Research.

Chamberlain, A. Ray and Sorrentino, Carl T. 1991. I-70 through Glenwood Canyon: "Showcase" public architecture of the 20th century. *AASHTO Quarterly Magazine* 70:4-6.

Christensen, Arthur G. and Jackson, Alvin N. 1969. Problems of relocation in a major city: Activities and achievements in Baltimore, Maryland. *Highway Research Record* 277:1-8.

Clayton, Christopher. 1977. Interstate population migration process and structure in the United States, 1935 to 1970. *The Professional Geographer* 29:177-81.

Cohen, Yehoshua S. and Berry, Brian J. 1975. *Spatial components of manufacturing change*. University of Chicago, Department of Geography 172.

Cromley, Robert G. and Leinbach, Thomas R. 1981. The pattern and impact of the filter down process in nonmetropolitan Kentucky. *Economic Geography* 57:208-24.

Crowds hail dedication of I-70's last leg. 1992. *The Toledo Blade*, October 15, p. 10.

Dearing, Charles L. and Owen, Wilfred. 1949. *National transportation policy*. Washington: Brookings Institute.

Drug mules shift to turnpike. 1993. *The Toledo Blade*, January 18, p. 9.

Dunn, James A. 1981. *Miles to go: European and American transportation policies*. Cambridge: MIT Press.

Eno Transportation Foundation, Inc. 1987. *Commuting in America*. Westport, CT: Eno Transportation Foundation, Inc.

_____. 1991. *Transportation in America*, 9th edition. Waldorf, MD: Eno Transportation Foundation, Inc.

Erickson, Rodney A. 1981. Corporations, branch plants, and employment stability in nonmetropolitan areas. In *Industrial location and regional systems*, ed. J. Rees; G. J. Hewings; and H. A. Stafford, p. 135-53. New York: J.F. Bergin Publishers, Inc.

_____, and Gentry, Marylynn. 1985. Suburban nucleations. *GeographicalReview* 75:19-31.

Eyerly, R. W.; Twark, Richard D. ; and Downing, R. H. 1987. Interstate highway system: reshaping the non-urban areas of Pennsylvania. *Transportation Research Record* 1125:1-7.

Finn, Edwin A. 1987. Cruising into the 21st century. *Forbes* August 24:80-3.

Fisher, James S. and Mitchelson, Ron L. 1981. Extended and internal commuting in the transformation of the intermetropolitan periphery. *Economic Geography* 57:189-207.

Flink, James J. 1990. *The automobile age.* Cambridge: MIT Press.

Friedlaender, Anne F. 1965. *The interstate highway system, a study in public investment.* Amsterdam: North Holland Publishing Company.

Fuguitt, Glen V. and Beale, Calvin L. 1976. *Population change in nonmetropolitan cities and towns.* Economic Development Division, Economic Research Service, Agricultural Economic Report No. 323. Washington: U.S. Department of Agriculture.

Garrison, William L. 1974. Connectivity of the interstate highway system. In *Transportation geography,* ed. M. Eliot Hurst, p. 81-90. New York: McGraw-Hill, Inc.

_____; Berry, Brian J.; Marble, D. F.; Nystuen, J. D.; and Morrill, R. L. 1959. *Studies of highway development and geographic change.* Seattle: University of Washington Press.

Hamilton, Roger, H. 1988. Identification and ranking of environmental impacts associated with the United States interstate highway system. *Transportation Research Record* 1166:1-8.

Hebert, Richard. 1972. *Highways to nowhere: The politics of city transportation.* Indianapolis: Bobbs-Merrill Company, Inc.

Heymann, Hans, Jr. 1965. The objectives of transportation. In *Transport investment and economic development,* ed. G. Fromm, p. 18-33. Washington: Brookings Institute.

Irwin, Larry L.; Mason, Mark L.; and Ward, A. Lorin. 1981. Lead compounds in mule deer and vegetation along I-80, southeastern Wyoming.*Transportation Research Record* 805:3-5.

Kelley, Ben. 1971. *The pavers and the paved.* New York: Donald W. Brown, Inc.

Larson, Theodore D. 1964. The motor road: forerunner of the universal city. *Traffic Quarterly* 18:459-90.

Lass, William. 1956. *Freedom of the American road.* Detroit: Ford Motor Company.

Lichter, Daniel T. and Fuguitt, Glen V. 1980. Demographic response to transportation innovation: The case of the interstate highway. *Social Forces* 59:492-512.

Lockwood, Kessler & Bartlett, Inc. 1992 black and white photo of the Alexander Hamilton Interchange on the George Washington Bridge in New York City, NY. Syosset, NY: Lockwood, Kessler & Bartlett, Inc.

Lowry, Ira S. 1988. Planning for urban sprawl. In *A look ahead: Year 2020,* ed. Transportation Research Board, p. 275-312. Washington: Transportation Research Board, National Research Council.

Lupo, Alan; Colcord, Frank; and Fowler, Edmund P. 1971. *Rites of way: The politics of transportation in Boston and the U.S. city.* Boston: Little, Brown and Company.

Mattes, Merrill J. 1983. *Scotts Bluff.* National Park Service Historical Handbook Series No. 28, Washington: U. S. Government Printing Office.

McKelvey, Blake 1973. *American urbanization: A comparative history.* Glenview, IL: Foresman and Company.

Mertins, Herman 1972. *National transportation policy in transition.* Lexington: Lexington Books.

Mitchell, Robert L. 1958. Toll roads and the interstate system. *Traffic Quarterly* 12:322-33.

Mitchelson, Ron L., and Fisher, James S. 1987a. Long-distance commuting and population change in Georgia. *Growth and Change* 18:44-65.

_____. 1987b. Long distance commuting and income change in the towns of upstate New York. *Economic Geography* 63:48-65.

_____. 1988. *Population and income growth multipliers for extended and localized commuting.* Athens, GA: U.S. Department of Commerce, Economic Development Administration.

Moon, Henry. 1986a. *Modeling land use change around interstate highway interchanges in nonmetropolitan areas: A multivariate statistical analysis.* Ph.D. thesis, Department of Geography, The University of Kentucky, Lexington, KY.

_____. 1986b. Regional variations in development along interstate highways in nonmetropolitan Kentucky. *Proceedings of the Third Conference on Appalachian Geography* 3:1-10.

_____. 1987a. Interstate highway inter-changes as instigators of nonmetropolitan development. *Transportation Research Record* 1125:8-14.

_____. 1987b. Interstate highway inter-changes reshape rural communities. *Rural Development Perspectives* 4:35-9.

_____. 1988. Modelling land use change around nonurban interstate highway interchanges. *Land Use Policy* 5:394-407.

_____. 1989. Interstate villages as urban places. *Small Town* 19:4-14.

_____. 1992. The interstate highway system. In *Snapshots of America*, ed. D. G. Janelle, p. 425-427. New York: Guilford Publications.

Nelson, James C. 1973. A critique of governmental intervention in transport. In *Perspectives on regional transportation planning*, ed. Joseph S. DeSalvo, p. 229-90. Lexington, MA: Lexington Books.

Newman, P. W. and Kenworthy, J. R. 1989. *Cities and automobile dependence: A sourcebook.* Aldershot, Hants, England: Gower Technical.

Norris, Darrell A. 1987. Interstate highway exit morphology: Nonmetro-politan exit commerce on I-75. *The Professional Geographer* 39:23-32.

Organisation for Economic Co-Operation and Development (OECD). 1988. *Cities and transport.* Paris: OECD.

Parker, Elizabeth. 1991. Major proposals to restructure the highway program. *Transportation Quarterly* 45:55-66.

Patrick, K. 1991. Transportation corridor: Examining the development and spatial structure of linear cultural landscapes. Paper presented at the Annual Meeting of the Association of American Geographers, Miami, FL.

Patton, Phil. 1986. *Open road: A celebration of the American highway.* New York: Simon and Schuster.

Payne-Maxie Consultants and Blayney-Dyett, Urban and Regional Planners. 1980. *The land use and urban development impacts of beltways: Executive summary.* Washington: U.S. Printing Office.

Plous, F. K., Jr. 1993. Refreshing ISTEA. *Planning* 59:9-12.

Prentiss, Louis W. 1962. Highway construction as an anti-recession activity. *Traffic Quarterly* 16:351-56.

Reno, Arlee T. 1988. Personal mobility in the United States. In *A look ahead: Year 2020*, ed. Transportation Research Board, pp. 369-93. Washington: Transportation Research Board, National Research Council.

Road Information Program, The. 1990. The federal highway trust fund. *Transportation Quarterly* 41:23-35.

Robertson, K. A. 1980. The impact of transportation on the central business district. *Traffic Quarterly* 34:523-38.

Rose, Mark H. 1990. *Interstate: Express highway politics 1939-1989.* Knoxville: University of Tennessee Press.

Rubenstein, James M. 1990. Japanese motor vehicle producers in the U.S.A.: Where and why. *Focus* 40:7-11.

_____. 1992. America's "just-in-time" highways: I-65 and I-75. In *Snap-shots of America,* D. G. Janelle, p. 432-35. New York: Guilford Publications.

Small, K. A.; Winston, C.; and Evans, C. A. 1989. *Road work: A new highway pricing and investment policy.* Washington: Brookings Institute.

Smerk, George M. 1965. *Urban trans-portation: The federal role.* Bloomington: Indiana University Press.

Smith, W. R. and Selwood, D. 1983. Office location and the density-distance relationship. *Urban Geography* 4:302-16.

Steptoe, Roosevelt and Thornton, Clarence. 1986. Differential influence of an interstate highway on the growth and development of low-income minority communities. *Transportation Research Record* 1074:60-68.

Taaffe, Edward J. and Gauthier, Howard L. 1973. *Geography of transportation.* Englewood Cliffs: Prentice-Hall, Inc.

Teachers Insurance and Annuity Association (TIAA) - College Retirement Equities Fund. 1992. Mall of America: An investment in innovation. *The participant.* New York: TIAA-CREF.

Thiel, Floyd I. 1962. Social effects of modern highway transportation. *Public Roads* 32:1-10.

Traffic deaths hit 30-year low. 1992. *The Toledo Blade,* December 30.

Twark, Richard D. 1967. *A predictive model of economic development at non-urban interchange sites on Pennsylvania interstate highways*. Ph.D. thesis, Department of Business Logistics, Pennsylvania State University, University Park, PA.

Ullman, Edward. 1956. The role of transportation and the bases for interaction. In *Man's role in changing the face of the earth*, ed. W. L. Thomas, p. 862-80. Chicago: The University of Chicago.

University of Iowa Public Policy Center, The, in conjunction with The Midwest Transportation Center. 1990. *Road investments to foster local economic development*. Iowa City, IA: University of Iowa.

U.S. Department of Commerce, Bureau of Public Roads (USDOC BPR). 1957-1966a. *Annual press releases*. Washington: U.S. Department of Commerce.

_____. 1959-1962b. *Highway statistics*. Washington: U.S. Government Printing Office.

U.S. Congress. 1887. *An act to regulate commerce*. Chap. 104, 49th Cong., 2nd Session, Washington.

_____. 1916. *Federal aid in the construction of rural post roads*. Senate Report No. 250., 64th Cong., 1st Session, Washington.

_____. 1928. *Federal highway act, amendment*. Chap. 660, 70th Cong., 1st Session, Washington.

_____. 1932. *Emergency relief and construction act*. July 21, Chap. 520, 47 Stat. 709.

_____. 1933. *National industrial recovery act*. June 16, Chap. 90, 48 Stat. 195.

_____. 1939. *Toll roads and free roads*. House Document No. 272, 76th Cong., 1st Session, Washington.

_____. 1940. *Transportation act of 1940*. Sept. 18, Chap. 722, 54 Stat. 898.

_____. 1949. *Highway needs of the national defense.* House Document No.249, 81st Cong., 1st Session, Washington.

_____. 1956a. *Federal highway and highway revenue acts of 1956.* House Report No. 2022, 84th Cong., 2nd Session, Washington.

_____. 1956b. *Federal highway and highway revenue acts of 1956.* House Report No. 2436, 84th Cong., 2nd Session, Washington.

_____. 1991. *Intermodal surface transportation efficiency act of 1991.* Dec. 18, Public Law 102-240, 105 Stat. 1914.

U.S. Department of Transportation (USDOT). 1992. *A summary: Intermodal surface transportation efficiency act of 1991.* Washington: U.S. Department of Transportation.

U.S. Department of Transportation, Federal Highway Administration (USDOT FHWA). 1967-1992a. *Annual press releases.* Washington: Federal Highway Administration.

_____. 1970-1991b. *Highway statistics.* Washington: Government Printing Office.

_____. 1970c. *Benefits of interstate highways.* Washington: Federal Highway Administration.

_____. 1976c. *America's highways.* Washington: U.S. Government Printing Office.

_____. 1976d. *Social and economic effects of highways.* Washington: Federal Highway Administration.

_____. 1984c. *America on the move!* Washington: Federal Highway Administration.

_____. 1985c. *Highway statistics: Summary to 1985.* Washington: Government Printing Office.

U.S. Department of Transportation, National Highway Traffic Safety Administration (USDOT NHTSA). 1991. *Fatal* *accident reporting system 1990*. Washington: National Highway Traffic Safety Administration.

Vance, James E. 1986. *Capturing the horizon: The historical geography of transportation*. New York: Harper and Row, Publishers.

Walters, David C. 1992. Repairing America. *The Christian Science Monitor* September 28, 9-12.

Wayman, Neva and Hassell, Sue. 1992. *Arkansas state and county economic data*. Little Rock: Arkansas Institute for Economic Advancement, University of Arkansas.

Weiner, E. 1986. *Urban transportation planning in the United States*. Washington: U.S. Department of Transportation.

Wheat, Leonard F. 1969. The effect of modern highways on urban manufacturing growth. *Highway Research Record* 277:9-24.

Wilder, Margaret G. 1985. Site and situation determinants of land use change: an empirical example. *Economic Geography* 61:332-44.

Wilson, F. R.; Graham, G. M.; and Aboul-Ela, Mohamed. 1985. Highway investment as a regional development policy tool. *Transportation Research Record* 1046:10-14.

Wolfe, Roy I. 1963. *Transportation and politics*. Searchlight Book #18, Princeton: D. Van Nostrand Company, Inc.

Worden, Don. 1992a. Interview with author. Little America, WY. August.

_____. 1992b. *Photo of operations, Wyoming (1992)*. Little America, WY: Little America, Inc.

Zemotel, Linda M.; Bullen, A. G; and Hummon, Norman P. 1987. Issues in planning for the transportation needs of advanced technology firms. *Transportation Research Record* 1125:15-20.